Acta Physica Austriaca

Supplementum XII

1974

Springer-Verlag
Wien New York

Formulae and Results in Weak Interactions

H. Pietschmann, Vienna

1974

Springer-Verlag
Wien New York

Prof. Dr. H. PIETSCHMANN
Institute for Theoretical Physics
University of Vienna, Austria,
and
Institute for High Energy Physics
Austrian Academy of Sciences

Printed in Austria

Library of Congress Cataloging in Publication Data

Pietschmann, Herbert, 1936-
Formulae and results in weak interactions.

(Acta physica Austriaca. Supplementum 12)
Bibliography: p.
1. Weak interactions (Nuclear physics) I. Title.
II. Series.
QC794.8.W4P53 539.7'54 74-10767

ISBN 3-211-81258-X Springer-Verlag Wien-New York
ISBN 0-387-81258-X Springer-Verlag New York-Wien

Foreword

It is a great pleasure for me to write some introductory remarks to this supplement volume to Acta Physica Austriaca. The reason for this is that the author not only is a competent expert in the field of weak interactions but also a longstanding personal friend of mine. Since the topic presented here is closely related to the subjects covered in our annual "Internationale Universitätswochen für Kernphysik der Universität Graz" at Schladming, where the author has lectured already quite a few times, I readily agreed to publish this compendium in our series, although it is for the first time that a single author's work is published in this form. Another factor in this decision was that the presentation of the material shows the same spirit as the proceedings of our conferences: it should enable students and researchers alike to find their way through the various formalisms of modern elementary particle physics.

One of the main advantages of this work is its coherence. It is not a collection taken from various papers in the literature; rather, the author has taken pains in computing and rechecking all the equations himself with some help from his research associates. Therefore one can assume that this collection will be of real use and help in the tedious day-to-day work of weak interaction physicists. In particular for experimentalists it should be extremely useful to find a major part of weak interaction physics covered in such a consistent way so that formulae can be taken without further worry about numerical factors.

We hope that this collection will meet its success. If that is realized, a further edition—possibly enriched and certainly updated—may be envisaged. Therefore comments and criticisms are invited in order to help improve a possible second edition.

P. Urban

Preface

Weak interaction phenomena occur in a wide variety of physical processes and a huge amount of empirical facts has been accumulated both in the field of nuclear physics and elementary particle physics. This is complemented by a theory which has been improved steadily since it was born in the late fifties. Thus we face a situation where we can compute most of the (low energy) processes and phenomena due to weak interactions and rely on the results. Yet the theory is not in its final version which is clear from the persistent divergences of higher order calculation. Therefore, the effort in improving the theory still further must not relax. To achieve the final goal of an ultimate theory of Weak Interactions, surely a difficult and cumbersome path has to be overcome. The present collection of formulae and results in Weak Interaction physics is meant as a "walking stick" to facilitate the advance on this path.

Rather than collecting all existing formulae and equations from the literature (which would be near to impossible anyway) we have put emphasis on the usefulness and handiness of this collection. We have thus only included results which we have personally recalculated and selected those which are of particular use and help according to our personal opinion. It is clear from this, that the collection represents to some extent the personal working method of the author. This does not mean that other methods or conventions may not be equivalent or even superior; however, we believe that this collection can only be useful—if at all—when it does not radiate the air of an impersonal dictionary or handbook.

Naturally, the whole research group of the author has contributed to this collection by its mere existence. However, some of the associates who have helped in checking and even contributing some of the equations deserve mentioning in the preface rather than in the appendix: Thanks are due to Drs. G. Ecker, H. Grosse, R. Kögerler, H. Kühnelt, and H. Stremnitzer for their help.

Vienna, July 1974 H. Pietschmann

Contents

I. Introduction 1
II. Some General Mathematical Equations 5
1. Operator-Identities 5
2. The Metric and the Anti-Symmetric Tensor 5
III. The Dirac-Algebra 7
1. The Dirac-Equation and γ-Matrices 7
2. Hermiticity Properties and Matrix $\mathscr{C}$ 7
3. The Dirac-Algebra and Fierz-Transformations 8
4. Projection Operators and Traces 9
IV. Field Operators and Invariant Functions 12
1. Field Operators 12
2. Normalization of States 12
3. Invariant Functions 12
V. S-Matrix, Cross-Sections and Life-Times 16
VI. Phase Space Integrals 17
1. General Formulae 17
2. Two-Particle Integrals 18
3. Three-Particle Integrals 19
4. n-Particle Integrals 21
VII. Weak Interaction Hamiltonian and S-Operator 22
1. Standard V-A Theory 22
2. Unified Models 23
VIII. Discrete Transformations 25
1. Space Reflections 25
2. Charge Conjugation 25
3. Time Reflection 26
4. G-Parity 26
IX. Muon-Decay 27
1. The Matrix Elements 27
2. The Electron Spectrum 27
3. Effects of the IVB 28
X. Semi-Leptonic Baryon Decay 29
1. List of Decay-Processes and Selection Rules 29
2. Matrix-Element and Decay Spectra 30
3. Restrictions on the Form Factors 33
4. The Cabibbo Current 34
5. Polarization Effects 35
XI. Semi-Leptonic Meson Decays 36
1. List of Decay Processes and Selection Rules 36
2. P_{l2}-Decays 37
3. P_{l3}-Decays 38

XII. Current-Algebra 40
1. General Relation 40
2. The Basic Commutators 40
3. The Goldberger-Treiman Relation 42
4. The Adler-Weisberger Relation 43
5. The Mathur-Okubo-Pandit-Callan-Treiman Relation 43
XIII. Hadronic Hyperon Decay 44
1. Matrix Element and Decay Rate 44
2. Asymmetry Parameters 45
3. The $\Delta I = \frac{1}{2}$ Rule 45
XIV. Quasi-Elastic Neutrino Scattering 47
1. Neutrino-Lepton Scattering 47
2. Neutrino-Baryon Scattering 48
XV. Deep Inelastic Neutrino Scattering 50
1. Kinematics 50
2. Laboratory Cross-Section and Structure Functions 50
3. Scaling Behaviour 52
4. Crossing, Isospin and Positivity Relations 52
5. Helicity Cross-Sections 53
6. Total Cross-Sections 54
7. Spectrum Independent Quantities 55
8. Influence of the W-Boson 55
XVI. Decay of the W-Boson 57
1. General Two-Body Decay 57
2. Decay Rates 57
Appendix A: Other Forms of the Interaction 59
1. The Non-Local Interaction 59
2. The Renormalizable Scalar Theory 59
3. Limits on Additional S, P, T Interactions 60
Appendix B: Comparison between Different Metrics 61
References 62
Sources 63

I. Introduction

In preparing a collection of formulae and results in a field with such a wide span as weak interactions, quite a number of decisions and selections has to be made. Each of these decisions and selections in itself may be argued. It is quite certain, that the reader will soon run into a formulation or a choice of convention which he finds to be impractical in his own opinion or where he has rather strong feelings towards a different choice. Is it not interesting, that in a field like elementary particle physics, where the matter itself is rather independent of personal views of the researcher and discoverer, much of the strong personal feelings is diverted into the completely neutral field of conventions, definitions, ways of formulating and the like? Though it is our opinion, that those who cannot take part in this harmless quarrel of "time-like versus space-like metric," to give but one example, have to be pitied, we want to emphasize very strongly that we have avoided to let our decisions and selections be influenced by any strong personal feeling. As far as this collection is concerned, *any* choice is equally good provided that there is maximal internal consistency.

What, then, are the motives for our choices and selections? The answer is clearly the aim at maximal consistency and minimal number of mistakes. Therefore, the choices are mainly driven by the history of the author's works and that of his research group. For example, the metric and Dirac algebra conventions coincide with that of the wide-spread book "Relativistic Quantum Mechanics" by J. D. Bjorkén and S. D. Drell [1], with one exception: The sign of γ_5 is opposite. Would it not be preferable—for the sake of unification—to use the same sign also for γ_5? The answer is, in principle yes! However, the author had used this convention before the book of Bjorkén and Drell appeared and the relevant formulae were thoroughly checked. A change in so many equations had introduced a factor of uncertainty which is better avoided. The price is the different definition of γ_5.

From this one example, it should be apparent that it is a purely pragmatic point of view which led to our choices. Since we are discussing metric and Dirac algebra, a further word of reconciliation is in order.

There is generally no doubt that the development of weak interaction physics was greatly influenced by the "predictor" of the neutrino: Wolfgang Pauli. It is also well-known that he preferred the metric with imaginary time-component and that this metric has also been used by other famous investigators who lateron added considerable steps to the struggle of shaping the theory of weak interactions into its ultimate form. However, there are also those who prefer the metric used in our collection and nobody would gladly like to weigh the contributions of either one "set of theorists" against the other. Therefore, we have to live with the fact that more than one convention is in use. It happens repeatedly, that co-authors use different metric in their computations. In order to facilitate their task for unifying their results to write a publication, we have added a "dictionary" in Appendix B, translating some of the most fundamental equations from our metric into the other one.

An almost equally "important" choice is the normalization of states and the definition of the T-matrix element. For reasons very similar to the ones just explained, we have chosen the non-invariant normalization of states. Advantages and disadvantages are much discussed and we do not want to repeat the arguments. Again, we have no strong feeling for either choice, except that all of the equations have been derived and checked in the convention chosen.

This brings us to another point: completeness! Naturally, a collection like this one is only useful if it covers a field completely. We have therefore included practically all parts of weak interaction with the exception of CP-violation. The latter effects have become more and more a subject of its own and we did not feel it useful or necessary to include them here.

The reader will find that the chapter on semi-leptonic meson decays does not contain K_{l4} decays. The reason is simply that neither the author nor one of his associates has ever worked on K_{l4} decays with any appreciable depth. It is one of the principles of this collection that no equation is just taken from the literature. The danger of errors entering this way is far too big! All of the equations have been computed and checked by the author; a few exceptions have been thoroughly checked by the associates mentioned in the preface. The advantage bought in this way is obviously a maximum trust in mutual consistency; on the other hand, no excuse can be made for errors other than to blame the author!

Except for the minimum necessary hints and verbal definitions, there is no text in the collection proper. We felt that a maximum concentration is helpful for using the collection in actual work. The reader—or rather the user—should read the introduction but afterwards he should not be bothered by too many words which often blur the picture. The user is advised to add as many remarks as possible in his own writing next to and between the equations. In this way, the collection may soon acquire some personal touch of the user's spirit which seems so necessary for an optimal success.

For the same reason, the number of references is rather limited. Clearly, some references to experimental results were necessary. Above that, only occasionally did we feel it proper to give a hint to original papers. The user is assumed to have a working knowledge of at least part of weak interactions, otherwise this collection will remain unintelligible. Therefore, the user is supposed to be familiar with names such as "Adler-Weisberger relation" and the like and reference to the original papers seemed not appropriate. On the other hand, we did add a list of "sources" so that one or the other minor point as to the meaning of a symbol or so can rapidly be cleared up. However, we would like to warn that factors can be different in the source-papers from this collection due to some updating of normalizations over the years.

It is quite obvious from the table of content, that areas on the border of the field or those where things are not well established have been left out. Examples are hadronic meson decays or resonance production by neutrinos. No further excuse is necessary. A more delicate point is the field of higher orders! From all the renormalizable gauge models the Salam-Weinberg model has been selected and extensively treated. This seems to be justified because this model is at least the most favorable one from the experimental point of view. We have not touched the subject of higher orders. Had we decided to do—and it certainly was considered—the whole subject of field theory including Feynman integrals and the like had blown up this collection far beyond the scope of weak interaction physics. If there is enough interest, we may one day extend this collection to contain these fields also.

Before we relief the reader from so much text to concentrated formulae, a few pragmatic hints are in order. In the chapter on some general mathematical equations, we have included only those formulae which are either of direct need in weak interactions or are often difficult to find. Very important mathematical tools such as particular integrals or power series expansions can easily be found in the rich literature of mathematical tables.

Whenever we use the term "log" we mean the natural logarithm (base e).

In the tables for particle decays, the anti-particle decays have not been listed separately. They can be trivially obtained by substituting the relevant anti-particle for each particle. Consequently, only π^+ and K^+ decays are listed because π^- and K^- decays are obtained as anti-particle decays.

We have avoided referring back to previous chapters for definitions of frequently occurring abbreviations. Instead, we simply repeat them whenever it seems appropriate. However, within one and the same chapter, no repetitions have been made. The user will often find it useful to indicate one or another definition, which he is not familiar with, on several places in his own writing.

It is almost self-evident that we use natural units, i.e.

$$\hbar = c = 1.$$

Consequently, the following relations can be used to express the remaining dimensional quantities through each other:

$$1\ (\mathrm{MeV})^{-1} = 6.5822 \cdot 10^{-22}\ \mathrm{sec}$$
$$= 1.97329 \cdot 10^{-11}\ \mathrm{cm}.$$

The brilliance and exactness of a part of natural science such as weak interactions too often tends to induce one to forget the tedious path which led to our deep knowledge. In other words, physicists too often lack the proper feeling for the historical dimension of their beautiful science. Therefore, we want to conclude the introduction with a reminder to the most important historical events in weak interactions as well as the larger field in which it is embedded: Elementary particle physics.

Historical Table

Radioactivity	1896	BECQUEREL
Electron	1897	J. J. THOMSON
Quantum Theory	1900	PLANCK
Relativity	1905	EINSTEIN
Nucleus	1911	RUTHERFORD
Atomic Model	1913	BOHR
Wave Equation	1926	SCHRÖDINGER
Uncertainty Principle	1927	HEISENBERG
Neutron	1932	CHADWICK
Positron	1932	ANDERSON
Neutrino-Hypothesis	1933	PAULI
Theory of β-Decay	1934	FERMI
Artificial Radioactivity	1934	CURIE and JOLIOT
Muon	1936	ANDERSON, NEDDERMEYER, STEVENSON, STREET
Pi-Meson	1946	POWELL
Parity Violation	1956	LEE and YANG
(Anti) Neutrino	1956	REINES and COWAN
V-A Theory	1958	FEYNMAN, GELL-MANN, MARSHAK, SUDARSHAN
Unitary Symmetry	1961	GELL-MANN
Two Neutrinos	1962	DANBY, GAILLARD, GOULIANOS, LEDERMAN, MISTRY, SCHWARTZ, STEINBERGER
Weak Interactions and SU(3)	1963	CABIBBO
CP-Violation	1964	FITCH, CRONIN, CHRISTENSON, TURLEY
Unified Gauge Theory	1967	SALAM, WARD, WEINBERG
Double β-Decay	1968	KIRSTEN, SCHAEFFER, NORTON, STOENNER
Neutral Weak Currents	1973	Collaboration: Aachen, Brussels, CERN, Paris, Milano, Orsay, London

II. Some General Mathematical Equations

1. Operator-Identities

$$\exp(iA)\,B\exp(-\,iA) = B + i[A, B] + \frac{i^2}{2!}[A, [A, B]] + \cdots$$

$$[A, \exp(-\,\beta B)] = -\exp(-\,\beta B)\int_0^\beta d\lambda \exp(\lambda B)\,[A, B]\exp(-\,\lambda B)$$

$$e^A e^B = \exp\{A + B + \tfrac{1}{2}[A, B] + \tfrac{1}{12}[[A, B], B] + \tfrac{1}{12}[[A, B], A] + \cdots\}$$

$$\operatorname{Det}|1 + A| = \exp\left\{\operatorname{Tr}\sum_{n=1}^{\infty}\frac{A^n}{n}\right\}$$

$$\begin{aligned}[AB, C] &= A[B, C] + [A, C]B \\ &= A\{B, C\} - \{A, C\}B\end{aligned}$$

where

$$\{A, B\} = AB + BA$$

$$\begin{aligned}[AB, CD] = {} & A\{B, C\}D - C\{A, D\}B \\ & + CA\{B, D\} - \{A, C\}BD\end{aligned}$$

2. The Metric and the Anti-Symmetric Tensor

Def.:

$$\varepsilon_{\alpha\beta\gamma\delta} = \begin{cases} 1 & \text{even permutations of } 0, 1, 2, 3 \\ 0 \quad \text{for} & 2 \text{ or more indices equal} \\ -\,1 & \text{odd permutations of } 0, 1, 2, 3 \end{cases}$$

$$g_{\mu\nu} = \begin{pmatrix} 1 & 0 & 0 & 0 \\ 0 & -\,1 & 0 & 0 \\ 0 & 0 & -\,1 & 0 \\ 0 & 0 & 0 & -\,1 \end{pmatrix}$$

$$x^\mu = (x^0, \mathbf{x}), \quad \mu, \nu, \ldots = 0, 1, 2, 3, \quad k, l, \ldots = 1, 2, 3$$

$$x_\mu = g_{\mu\nu} x^\nu$$

$$(ab) = a_\mu b^\mu = a^0 b^0 - \mathbf{ab}$$

$$\not{a} = a_\mu \gamma^\mu$$

$$\partial_\mu = \frac{\partial}{\partial x^\mu}$$

$$\varepsilon_{\alpha\beta\gamma\delta}\varepsilon^{\alpha\beta\gamma\delta} = -24$$

$$\varepsilon_{\alpha\beta\gamma\mu}\varepsilon^{\alpha\beta\gamma\nu} = -6\delta_\mu{}^\nu$$

$$\varepsilon_{\alpha\beta\mu\nu}\varepsilon^{\alpha\beta\rho\sigma} = -2(\delta_\mu{}^\rho\delta_\nu{}^\sigma - \delta_\mu{}^\sigma\delta_\nu{}^\rho)$$

$$\varepsilon_{\alpha\mu\nu\sigma}\varepsilon^{\alpha\lambda\rho\tau} = -\{\delta_\mu{}^\lambda\delta_\nu{}^\rho\delta_\sigma{}^\tau + \delta_\mu{}^\rho\delta_\nu{}^\tau\delta_\sigma{}^\lambda + \delta_\mu{}^\tau\delta_\nu{}^\lambda\delta_\sigma{}^\rho$$
$$- \delta_\mu{}^\lambda\delta_\nu{}^\tau\delta_\sigma{}^\rho - \delta_\nu{}^\rho\delta_\mu{}^\tau\delta_\sigma{}^\lambda - \delta_\sigma{}^\tau\delta_\mu{}^\rho\delta_\nu{}^\lambda\}$$

III. The Dirac-Algebra

1. The Dirac-Equation and γ-Matrices

$$\left(\frac{\partial}{\partial t} + \alpha_k \nabla_k + i\beta m\right) \psi(x) = 0$$

$$\beta = \begin{pmatrix} I & 0 \\ 0 & -I \end{pmatrix} \qquad \alpha_k = \begin{pmatrix} 0 & \sigma_k \\ \sigma_k & 0 \end{pmatrix}$$

$$\gamma_0 = \beta$$

$$\gamma^k = \beta\alpha_k$$

$$\gamma_5 = -i\gamma^0\gamma^1\gamma^2\gamma^3 = \frac{-i}{4!}\varepsilon_{\mu\nu\sigma\lambda}\gamma^\mu\gamma^\nu\gamma^\sigma\gamma^\lambda$$

This definition of γ_5 differs from that of BJORKÉN and DRELL [1] by a sign!

$$(i\gamma^\mu\partial_\mu - m)\psi(x) = 0$$

$$i\partial_\mu\bar{\psi}(x)\gamma^\mu + m\bar{\psi}(x) = 0$$

$$\gamma^\mu\gamma^\nu + \gamma^\nu\gamma^\mu = 2g^{\mu\nu}$$

$$\gamma^\mu\gamma_5 + \gamma_5\gamma^\mu = 0$$

$$\gamma_5{}^2 = 1$$

$$\sigma^{\mu\nu} = \frac{i}{2}(\gamma^\mu\gamma^\nu - \gamma^\nu\gamma^\mu) = i(\gamma^\mu\gamma^\nu - g^{\mu\nu})$$

$$\sigma^{\mu\nu}\sigma_{\mu\nu} = 12$$

Def.: $$A\!\!\!/ = A_\mu\gamma^\mu$$

2. Hermiticity Properties and Matrix $\mathscr{C}$

T = transpose

† = hermitian conj.

$*$ = complex conj.

$$\gamma_0\gamma_\mu\gamma_0 = \gamma_\mu{}^\dagger$$
$$\gamma_0\sigma_{\mu\nu}\gamma_0 = \sigma_{\mu\nu}{}^\dagger$$

The charge conjugation matrix $\mathscr{C}$ has the properties

$$\mathscr{C}^{-1}\gamma_\mu\mathscr{C} = -\gamma_\mu{}^T$$
$$\mathscr{C}^{-1}\gamma_5\mathscr{C} = \gamma_5{}^T$$
$$\bar{u}\mathscr{C}^T = v^T$$
$$\mathscr{C}^T = -\mathscr{C}$$

For definition, compare Sec. VIII.2.

3. The Dirac-Algebra and Fierz-Transformations

Let

$$\{\Gamma_i\} = \left\{1, \gamma^\mu, \frac{1}{\sqrt{2}}\sigma^{\mu\nu}, i\gamma^\mu\gamma_5, \gamma_5\right\}$$

then the Pauli-identities are:

$$\delta_{\alpha\sigma}\delta_{\beta\tau} = \tfrac{1}{4}\sum_i (\Gamma_i)_{\tau\sigma}(\Gamma_i)_{\alpha\beta}$$

$$\varepsilon^{\lambda\mu\nu\sigma}\gamma_\mu\gamma_\nu\gamma_\sigma = 6i\gamma_5\gamma^\lambda$$

$$\gamma_5\sigma^{\alpha\beta} = \frac{i}{2}\varepsilon^{\mu\nu\alpha\beta}\sigma_{\mu\nu}$$

$$\gamma^\mu\gamma^\nu\gamma^\rho = g^{\mu\nu}\gamma^\rho - g^{\mu\rho}\gamma^\nu + g^{\nu\rho}\gamma^\mu + i\varepsilon^{\mu\nu\rho\sigma}\gamma_5\gamma_\sigma$$
$$\sigma^{\mu\nu}\gamma^\lambda = ig^{\nu\lambda}\gamma^\mu - ig^{\mu\lambda}\gamma^\nu + \varepsilon^{\mu\nu\lambda\alpha}\gamma_\alpha\gamma_5$$
$$\sigma_{\mu\nu}\gamma_\lambda - \gamma_\lambda\sigma_{\mu\nu} = 2i(g_{\lambda\nu}\gamma_\mu - g_{\lambda\mu}\gamma_\nu)$$
$$\sigma^{\mu\nu}\sigma^{\alpha\beta} = i\sigma^{\alpha\mu}g^{\beta\nu} - i\sigma^{\alpha\nu}g^{\beta\mu} + i\sigma^{\beta\nu}g^{\alpha\mu} - i\sigma^{\beta\mu}g^{\alpha\nu} +$$
$$+ g^{\alpha\mu}g^{\beta\nu} - g^{\alpha\nu}g^{\beta\mu} + i\varepsilon^{\mu\nu\alpha\beta}\gamma_5$$
$$\gamma^\mu\gamma_\mu = 4$$
$$\gamma^\mu A\!\!\!/\gamma_\mu = -2A\!\!\!/$$
$$\gamma^\mu A\!\!\!/B\!\!\!/\gamma_\mu = 4(AB)$$
$$\gamma^\mu A\!\!\!/B\!\!\!/C\!\!\!/\gamma_\mu = -2C\!\!\!/B\!\!\!/A\!\!\!/$$

Fierz transformations of the first kind (for second quantized fields!):

$$\bar{\psi}_a\Gamma_i\psi_b\bar{\psi}_c\Gamma_i\psi_d = -\sum_j \mathscr{F}_{ij}\bar{\psi}_a\Gamma_j\psi_d\bar{\psi}_c\Gamma_j\psi_b$$

$$\mathscr{F}_{ij} = \tfrac{1}{4}\begin{pmatrix} 1 & 1 & 1 & 1 & 1 \\ 4 & -2 & 0 & 2 & -4 \\ 6 & 0 & -2 & 0 & 6 \\ 4 & 2 & 0 & -2 & -4 \\ 1 & -1 & 1 & -1 & 1 \end{pmatrix}$$

furthermore, for the standard "V-A interaction":

$$\bar{\psi}_a\gamma^\lambda(1+\gamma_5)\psi_b\bar{\psi}_c\gamma_\lambda(1+\gamma_5)\psi_d = \bar{\psi}_a\gamma^\lambda(1+\gamma_5)\psi_d\bar{\psi}_c\gamma_\lambda(1+\gamma_5)\psi_b$$

Eigenvalues of the Fierz matrix ($-\mathscr{F}_{ij}$) are

$+1$ (3-fold degenerate)

-1 (2-fold degenerate)

Eigenvectors are:

symmetric: $2Q_s - Q_T - Q_A$

$Q_V - Q_A$

$Q_S - Q_V - Q_P$

antisymmetric: $2Q_S + Q_V + Q_A - 2Q_P$

$3Q_S + Q_T + 3Q_P$

where

$$Q_i = \bar{\psi}_a\Gamma_i\psi_b\bar{\psi}_c\Gamma_i\psi_d$$

Fierz transformations of the second kind (for second quantized fields!):

$$\bar{\psi}_a\gamma^\lambda(1+\gamma_5)\psi_b\bar{\psi}_c\gamma_\lambda(1+\gamma_5)\psi_d = -2\bar{\psi}_a(1-\gamma_5)\mathscr{C}\bar{\psi}_c{}^T\psi_d{}^T\mathscr{C}^{-1}(1+\gamma_5)\psi_b$$

4. Projection Operators and Traces

$$\sum_{\text{Spins}} |\bar{u}_f(p')\Gamma u_i(p)|^2 = \mathrm{Tr}\{\Gamma P_+(p)\Gamma' P_+(p')\}$$

with

$$\Gamma' = \gamma_0\Gamma^\dagger\gamma_0$$

For anti-particle spinors $v(p)$ or $\bar{v}(p)$ replace

$$P_+(p) \to -P_-(p)$$

with

$$P_\pm(p) = \frac{m \pm \not{p}}{2m}$$

The following normalization of Dirac-spinors is used ($i, j = 1, 2$):

$$\bar{u}_i(p)u_j(p) = \delta_{ij}$$
$$\bar{v}_i(p)v_j(p) = -\delta_{ij}$$
$$\bar{u}_i(p)v_j(p) = \bar{v}_i(p)u_j(p) = 0$$

also

$$u_i^\dagger(p)u_j(p) = \frac{E}{m}\delta_{ij} \quad \text{for} \quad m \neq 0$$

$$(\not{p} - m)u(p) = 0 \qquad (\not{p} + m)u(p) = 2mu(p)$$
$$(\not{p} + m)v(p) = 0 \qquad (\not{p} - m)v(p) = -2mv(p)$$

Spin projection operator:

$$S_n = \tfrac{1}{2}(1 - \gamma_5\gamma_\mu w^\mu)$$
$$w_\mu w^\mu = -1$$
$$w_\mu p^\mu = 0$$

and in the rest system

$$(w_\mu) \to (0, \mathbf{n})$$

$\mathbf{n}$ being the unit vector of the spin direction. In general, for

$$\sphericalangle(\mathbf{n}, \mathbf{p}) = \vartheta$$

$$\mathbf{w}^2 = \frac{E^2}{E^2\sin^2\vartheta + m^2\cos^2\vartheta} \qquad w_0^2 = \frac{\mathbf{p}^2\cos^2\vartheta}{E^2\sin^2\vartheta + m^2\cos^2\vartheta}$$

General equations for traces:

$$\mathrm{Tr}[\gamma_{\mu_1}\gamma_{\mu_2}\cdots\gamma_{\mu_n}] = 0 \quad \text{if} \quad n \text{ is odd}$$
$$\mathrm{Tr}\,\gamma_5 = \mathrm{Tr}(\gamma_\lambda\gamma_5) = \mathrm{Tr}(\gamma_\lambda\gamma_\nu\gamma_5) = \mathrm{Tr}(\gamma_\lambda\gamma_\nu\gamma_\sigma\gamma_5) = 0$$
$$\mathrm{Tr}\,AB = \mathrm{Tr}\,BA$$
$$\mathrm{Tr}(\gamma_\mu\gamma_\nu) = 4g_{\mu\nu}$$
$$\mathrm{Tr}(\gamma_\mu\gamma_\nu\gamma_\sigma\gamma_\lambda) = 4(g_{\mu\nu}g_{\sigma\lambda} + g_{\mu\lambda}g_{\nu\sigma} - g_{\mu\sigma}g_{\nu\lambda})$$
$$\mathrm{Tr}(\gamma_\mu\gamma_\nu\gamma_\sigma\gamma_\lambda\gamma_5) = -4i\varepsilon_{\mu\nu\sigma\lambda}$$

Let

$\Gamma = 1$	for	S
γ_5		P
$\sigma_{\mu\nu}$		$T_{\mu\nu}$
γ_λ		V_λ
$\gamma_\lambda\gamma_5$		A_λ

then the following traces hold:

Γ	Γ'	$\mathrm{Tr}[\Gamma(\not{p}_1 + m_A)\Gamma'(\not{p}_2 + m_B)]$
V_λ	V_μ	$2[p_\lambda p_\mu - q_\lambda q_\mu - g_{\lambda\mu}(\Delta^2 - q^2)]$
V_λ	A_μ	$2i\varepsilon_{\lambda\alpha\mu\beta}p^\alpha q^\beta$
V_λ	S	$2(p_\lambda\Sigma + q_\lambda\Delta)$
S	V_μ	$2(p_\mu\Sigma + q_\mu\Delta)$
S	S	$2(\Sigma^2 - q^2)$
A_λ	V_μ	$2i\varepsilon_{\lambda\alpha\mu\beta}p^\alpha q^\beta$
A_λ	A_μ	$2[p_\lambda p_\mu - q_\lambda q_\mu - g_{\lambda\mu}(\Sigma^2 - q^2)]$
A_λ	P	$2(p_\lambda\Delta + q_\lambda\Sigma)$
P	A_μ	$-2(p_\mu\Delta + q_\mu\Sigma)$
P	P	$2(q^2 - \Delta^2)$
V_λ	$T_{\sigma\mu}$	$2i[\Sigma(g_{\lambda\sigma}q_\mu - g_{\lambda\mu}q_\sigma) + \Delta(g_{\lambda\sigma}p_\mu - g_{\lambda\mu}p_\sigma)]$
A_λ	$T_{\sigma\mu}$	$2\Sigma\varepsilon_{\lambda\sigma\mu\alpha}p^\alpha + 2\Delta\varepsilon_{\lambda\sigma\mu\alpha}q^\alpha$
P	$T_{\sigma\mu}$	$2\varepsilon_{\sigma\alpha\mu\beta}p^\alpha q^\beta$
V_λ	$T_{\sigma\mu}\gamma_5$	$2\Sigma\varepsilon_{\lambda\sigma\mu\alpha}q^\alpha + 2\Delta\varepsilon_{\lambda\sigma\mu\alpha}p^\alpha$
A_λ	$T_{\sigma\mu}\gamma_5$	$2i[\Sigma(g_{\lambda\sigma}p_\mu - g_{\lambda\mu}p_\sigma) + \Delta(g_{\lambda\sigma}q_\mu - g_{\lambda\mu}q_\sigma)]$
P	$T_{\sigma\mu}\gamma_5$	$2i(p_\mu q_\sigma - q_\mu p_\sigma)$
S	A_μ	0
V_λ	P	0
S	P	0
A_λ	S	0
P	V_μ	0
P	S	0

where

$$p = p_2 + p_1$$
$$q = p_2 - p_1$$
$$\Sigma = m_A + m_B$$
$$\Delta = m_A - m_B$$

IV. Field Operators and Invariant Functions

1. Field Operators

The Boson field operator:

$$\phi(x) = \frac{1}{(2\pi)^{3/2}} \int d^3k \frac{1}{\sqrt{2k^0}} \{\alpha(k) \exp(-ikx) + \beta^\dagger(k) \exp(ikx)\}$$

The Dirac field operator:

$$\psi(x) = \frac{1}{(2\pi)^{3/2}} \int d^3p \sum_{s=1}^{2} \sqrt{\frac{m}{p_0}} [a_s(p) u_s(p) \exp(-ipx) + b_s^\dagger(p) v_s(p) \exp(ipx)]$$

$$\bar{\psi}(x) = \frac{1}{(2\pi)^{3/2}} \int d^3p \sum_{s=1}^{2} \sqrt{\frac{m}{p_0}} [a_s^\dagger(p) \bar{u}_s(p) \exp(ipx) + b_s(p) \bar{v}_s(p) \exp(-ipx)]$$

2. Normalization of States

$$\langle 0|0\rangle = 1$$

$$|k, s\rangle = a_s^\dagger(k)|0\rangle$$

$$\langle k', s'|k, s\rangle = \delta_{ss'}\delta^{(3)}(k' - k)$$

$$\vdots$$

3. Invariant Functions

$$\Delta_i = \frac{-1}{(2\pi)^4} \int_{C_i} d^4k \frac{\exp(-ikx)}{k^2 - m^2}$$

with the following paths C_i in the k^0-plane

Green's functions:

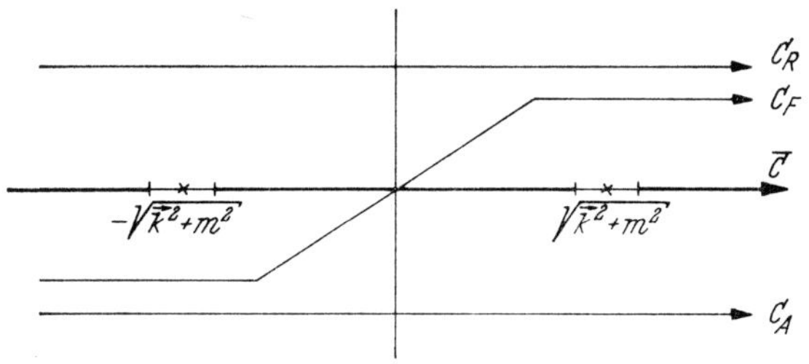

$$(\Box + m^2)\Delta_i(x) = \delta^{(4)}(x)$$

Homogeneous functions:

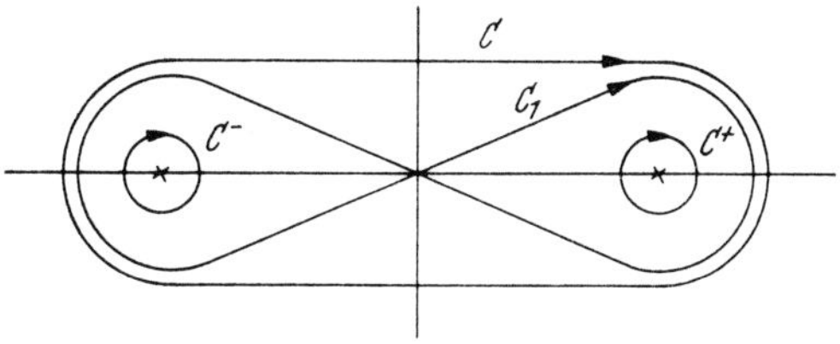

$$(\Box + m^2)\Delta_i(x) = 0$$

The following relations hold:

$$\bar{\Delta} = \tfrac{1}{2}(\Delta_R + \Delta_A)$$
$$\bar{\Delta}(x) = \bar{\Delta}(-x)$$
$$\bar{\Delta}^\dagger = \bar{\Delta}$$

$$\Delta_R(x) = \Delta_A(-x)$$
$$\Delta_R{}^\dagger = \Delta_R$$
$$\Delta_A{}^\dagger = \Delta_A$$

$$\Delta = \Delta^+ + \Delta^-$$
$$\Delta^1 = \Delta^+ - \Delta^-$$

$$\Delta^+(x) = -\Delta^-(x)$$
$$\Delta^1(x) = \Delta^1(-x)$$
$$\Delta(x) = -\Delta(-x)$$

$$\Delta^\dagger = \Delta$$

$$\Delta^{1\dagger} = -\Delta^1$$

$$\Delta^{-\dagger} = \Delta^+$$

$$\Delta_R = \bar{\Delta} + \tfrac{1}{2}\Delta$$

$$\Delta_A = \bar{\Delta} - \tfrac{1}{2}\Delta$$

$$\Delta_F = \bar{\Delta} + \tfrac{1}{2}\Delta^1$$

$$\Delta_F(x) = \Delta_F(-x)$$

$$\Delta_R(x) = 2\Theta(x^0)\bar{\Delta}(x)$$

$$\Delta_A(x) = 2\Theta(-x^0)\bar{\Delta}(x)$$

$$\Delta(x) = 2\varepsilon(x^0)\bar{\Delta}(x)$$

$$\Delta_F(x) = \Theta(x^0)\Delta^+(x) - \Theta(-x^0)\Delta^-(x)$$

$$\{\Delta^1; \Delta; \Delta^+; \Delta^-\} = \frac{i}{(2\pi)^3}\int d^4k \exp(-ikx)\,\delta(k^2 - m^2)\{1; \varepsilon(k^0); \Theta(k^0); -\Theta(-k^0)\}$$

$$\frac{\partial}{\partial x^0}\{\Delta^1; \Delta; \Delta^+; \Delta^-\}|_{x^0=0} = \{0; \delta^{(3)}(\mathbf{x}); \tfrac{1}{2}\delta^{(3)}(\mathbf{x}); \tfrac{1}{2}\delta^{(3)}(\mathbf{x})\}$$

$$\text{for } m = 0\text{: } \Delta \to D$$

$$\{\bar{D}; D; D_R; D_A\} = \frac{\delta(x^2)}{2\pi}\{\tfrac{1}{2}; \varepsilon(x^0); \Theta(x^0); \Theta(-x^0)\}$$

Invariant spinor functions:

$$S_i = \frac{1}{(2\pi)^4}\int_{C_i} d^4p\,\frac{\exp(-ipx)}{\not{p} - m}$$

$$S_i = -(i\not{\partial} + m)\Delta_i$$

Hankel function representation of Δ^+:

$$\Delta^+(x) = \frac{m}{8\pi\sqrt{z}} H_1^{(1)}(m\sqrt{z})$$

with

$$z = x_\mu x^\mu$$

Spectral representations for products of Δ^+ (colons represent partial derivatives):

Def.: $$w(a, b, c) = [a^2 + b^2 + c^2 - 2(ab + ac + bc)]^{1/2}$$
$$= [(a + b - c)^2 - 4ab]^{1/2}$$
$$= [a - (\sqrt{b} + \sqrt{c})^2]^{1/2}[a - (\sqrt{b} - \sqrt{c})^2]^{1/2}$$

$$\Delta^+(x, m^2)\Delta^+(x, M^2) = \frac{i}{16\pi^2}\int_{(m+M)^2}^{\infty}\frac{ds}{s}w(M^2, m^2, s)\Delta^+(x, s)$$

$$\Delta^+{}_{/\mu}(x, m^2)\Delta^+(x, M^2) = \frac{i}{32\pi^2}\int_{(m+M)^2}^{\infty}\frac{ds}{s^2}(s + m^2 - M^2)w(M^2, m^2, s)\Delta^+{}_{/\mu}(x, s)$$

$$\Delta^+{}_{/\mu\nu}(x, m^2)\Delta^+(x, M^2) = \frac{i}{48\pi^2}\int_{(m+M)^2}^{\infty}\frac{ds}{s^2}w(M^2, m^2, s)\left\{\tfrac{1}{4}[w(M^2, m^2, s)]^2 g_{\mu\nu}\right.$$
$$\left. + \left[\frac{1}{s}(s + m^2 - M^2)^2 - m^2\right]\partial_\mu\partial_\nu\right\}\Delta^+(x, s)$$

$$\Delta^+{}_{/\mu}(x, m^2)\Delta^+{}_{/\nu}(x, M^2) = \frac{-i}{48\pi^2}\int_{(m+M)^2}^{\infty}\frac{ds}{s^2}w(M^2, m^2, s)\left\{\tfrac{1}{4}[w(M^2, m^2, s)]^2 g_{\mu\nu}\right.$$
$$\left. - \frac{1}{2s}[s^2 + s(M^2 + m^2) - 2(M^2 - m^2)^2]\partial_\mu\partial_\nu\right\}\Delta^+(x, s)$$

$$\Delta^+{}_{/\nu\sigma}(x, m^2)\Delta^{\sigma+}{}_{/\mu}(x, M^2) = \frac{i}{48\pi^2}\int_{(m+M)^2}^{\infty}\frac{ds}{s^2}w(M^2, m^2, s)[s - (M^2 + m^2)]$$
$$\times\left\{\tfrac{1}{8}[w(M^2, m^2, s)]^2 g_{\mu\nu} - \frac{1}{4s}[s^2 + s(M^2 + m^2)\right.$$
$$\left. - 2(M^2 - m^2)^2]\partial_\mu\partial_\nu\right\}\Delta^+(x, s)$$

V. S-Matrix, Cross-Sections and Life-Times

$$\langle f|S|i\rangle = \delta_{fi} + i(2\pi)^4\delta^{(4)}(P_f - P_i)\cdot\mathcal{N}\cdot T_{fi}$$

$$\mathcal{N} = \prod_i \frac{1}{(2E_i)^{1/2}(2\pi)^{3/2}} \prod_f \frac{1}{(2E_f)^{1/2}(2\pi)^{3/2}}$$

Notice that T_{fi} contains a factor $\sqrt{2m}$ for each fermion!

Let n be the number of particles in the final state:

$$\sigma = \frac{1}{4E_aE_b}\frac{1}{v_{\text{in}}}\frac{1}{(2\pi)^{3n-4}}\int\frac{d^3p_1}{2E_1}\cdots\frac{d^3p_n}{2E_n}\,\delta^{(4)}(P_f - P_i)\sum_{\text{Spins}}{}'\,|T_{fi}|^2$$

where a and b are the initial particles, v_{in} is the flux and the prime on the sum recalls averaging over initial spins.

$$\frac{d\sigma}{dt_{ij}} = \frac{1}{4E_aE_b}\frac{1}{v_{\text{in}}}\frac{1}{(2\pi)^{3n-4}}\int\frac{d^3p_1}{2E_1}\cdots\frac{d^3p_n}{2E_n}\,\delta^{(4)}(P_f - P_i)$$

$$\times\,\delta[t_{ij} - (p_i - p_j)^2]\sum_{\text{Spins}}{}'\,|T_{fi}|^2$$

$$\Gamma(a\to n) = \frac{1}{\tau_{a,n}} = \frac{1}{2E_a}\frac{1}{(2\pi)^{3n-4}}\int\frac{d^3p_1}{2E_1}\cdots\frac{d^3p_n}{2E_n}\,\delta^{(4)}(P_f - P_i)\sum_{\text{Spins}}{}'\,|T_{fi}|^2$$

The power of π in a total cross-section or decay rate is

$$N_\pi = 3 - 2n$$

VI. Phase Space Integrals

1. General Formulae

$$\frac{1}{2E} = \int dp^0\, \Theta(p^0)\, \delta(p^2 - m^2) \quad \text{if} \quad E = {}_{+}\sqrt{\mathbf{p}^2 + m^2}$$

Def.: $$w(a, b, c) = [a^2 + b^2 + c^2 - 2(ab + ac + bc)]^{1/2}$$

$$= [(a + b - c)^2 - 4ab]^{1/2}$$

$$= [a - (\sqrt{b} + \sqrt{c})^2]^{1/2}[a - (\sqrt{b} - \sqrt{c})^2]^{1/2}$$

$$\int d^4q\, \Theta(q^0)\Theta(p^0 - q^0)\, \delta[(p - q)^2 - m^2]\, \delta(q^2 - M^2)$$

$$= \frac{\pi}{2} \Theta(p^0)\Theta[p^2 - (M + m)^2] \frac{1}{p^2} w(p^2, M^2, m^2)$$

$$\int d^4q\, \Theta(q^0)\Theta(p^0 - q^0)\, \delta[(p - q)^2 - m^2]\, \delta(q^2 - M^2) q^\mu$$

$$= \frac{\pi}{4} \Theta(p^0)\Theta[p^2 - (M + m)^2] \frac{1}{p^4} w(p^2, M^2, m^2) p^\mu (p^2 + M^2 - m^2)$$

$$\int d^4q\, \Theta(q^0)\Theta(p^0 - q^0)\, \delta[(p - q)^2 - m^2]\, \delta(q^2 - M^2) q^\mu q^\nu$$

$$= \frac{\pi}{6} \Theta(p^0)\Theta[p^2 - (M + m)^2] \frac{1}{p^4} w(p^2, M^2, m^2)$$

$$\times \left\{ \frac{p^\mu p^\nu}{p^2} [(p^2 + M^2 - m^2)^2 - p^2 M^2] - \frac{g^{\mu\nu}}{4} [w(p^2, M^2, m^2)]^2 \right\}$$

Let t be in the region such that

$$\left| \frac{P^2(t - M^2 - p^2) + (Pp)(P^2 + M^2 - m^2)}{w(P^2, M^2, m^2) w[P^2, p^2, (P - p)^2]} \right| \leqslant \tfrac{1}{2}$$

$$\int d^4q\, \Theta(q^0)\Theta(P^0 - q^0)\, \delta[(P-q)^2 - m^2]\, \delta(q^2 - M^2)\, \delta[(p-q)^2 - t]$$

$$= \frac{\pi}{2} \frac{1}{w[P^2, p^2, (P-p)^2]} \Theta(P^0)\Theta[P^2 - (M+m)^2]$$

$$\int d^4q\, \Theta(q^0)\Theta(P^0 - q^0)\, \delta[(P-q)^2 - m^2]\, \delta(q^2 - M^2)\, \delta[(p-q)^2 - t]q^\mu$$

$$= \frac{\pi\Theta(P^0)\Theta[P^2 - (M+m)^2]}{\{w[P^2, p^2, (P-p)^2]\}^3} \{P_\mu[(Pp)(p^2 + M^2 - t) - p^2(P^2 + M^2 - m^2)]$$

$$+ p_\mu[(Pp)(P^2 + M^2 - m^2) - P^2(p^2 + M^2 - t)]\}$$

2. Two-Particle Integrals

Let f be a function of one variable such that the integrals are defined.

$$\int \frac{d^3p_1}{2E_1} \frac{d^3p_2}{2E_2} \delta^{(4)}(P - p_1 - p_2) f(p_1p_2)$$

$$= \frac{\pi}{2} \frac{1}{P^2} w(P^2, m_1^2, m_2^2) f[\tfrac{1}{2}(P^2 - m_1^2 - m_2^2)]$$

$$m_i^2 = p_i^2 \qquad i = 1, 2$$

$$\int \frac{d^3p_1}{2E_1} \frac{d^3p_2}{2E_2} \delta^{(4)}(P - p_1 - p_2) f(p_1p_2) p_2^\mu$$

$$= \frac{\pi}{4} \frac{1}{P^4} w(P^2, m_1^2, m_2^2)[P^2 + m_2^2 - m_1^2] f[\tfrac{1}{2}(P^2 - m_1^2 - m_2^2)] P^\mu$$

$$\int \frac{d^3p_1}{2E_1} \frac{d^3p_2}{2E_2} \delta^{(4)}(P - p_1 - p_2) f(p_1p_2) p_2^\mu p_2^\nu$$

$$= \frac{\pi}{6} \frac{1}{P^4} w(P^2, m_1^2, m_2^2) f[\tfrac{1}{2}(P^2 - m_1^2 - m_2^2)]$$

$$\times \left\{ \frac{P^\mu P^\nu}{P^2} [(P^2 + m_2^2 - m_1^2)^2 - P^2 m_2^2] - \frac{g^{\mu\nu}}{4} [w(P^2, m_1^2, m_2^2)]^2 \right\}$$

$$\int \frac{d^3p_1}{2E_1} \frac{d^3p_2}{2E_2} \delta^{(4)}(P - p_1 - p_2) f(p_1p_2) p_1^\mu p_2^\nu$$

$$= \frac{\pi}{12} \frac{1}{P^4} w(P^2, m_1^2, m_2^2) f[\tfrac{1}{2}(P^2 - m_1^2 - m_2^2)]$$

$$\times \left\{ \frac{g^{\mu\nu}}{2} [w(P^2, m_1^2, m_2^2)]^2 + \frac{P^\mu P^\nu}{P^2} [P^4 + P^2(m_1^2 + m_2^2) - 2(m_1^2 - m_2^2)^2] \right\}$$

$$\int \frac{d^3p_1}{2E_1}\frac{d^3p_2}{2E_2}\,\delta^{(4)}(P-p_1-p_2)f(p_1p_2)p_2{}^\mu p_2{}^\nu p_2{}^\lambda$$

$$= \frac{\pi}{2}\frac{1}{P^6}\,w(P^2, m_1{}^2, m_2{}^2)f[\tfrac{1}{2}(P^2-m_1{}^2-m_2{}^2)]$$

$$\times\left\{\frac{P^\mu P^\nu P^\lambda}{2P^2}(P^2+m_2{}^2-m_1{}^2)[\tfrac{1}{2}(w(P^2, m_1{}^2, m_2{}^2))^2-P^2m_2{}^2]\right.$$

$$\left.-\tfrac{1}{24}(g^{\mu\nu}P^\lambda+g^{\mu\lambda}P^\nu+g^{\nu\lambda}P^\mu)(P^2+m_2{}^2-m_1{}^2)[w(P^2, m_1{}^2, m_2{}^2)]^2\right\}$$

$$\int \frac{d^3p_1}{2E_1}\frac{d^3p_2}{2E_2}\,\delta^{(4)}(P-p_1-p_2)f(p_1p_2)p_1{}^\mu p_2{}^\nu p_2{}^\lambda$$

$$= \frac{\pi}{2}\frac{1}{P^4}\,w(P^2, m_1{}^2, m_2{}^2)f[\tfrac{1}{2}(P^2-m_1{}^2-m_2{}^2)]$$

$$\times\left\{\frac{P^\mu P^\nu P^\lambda}{P^2}\left[\tfrac{1}{3}(P^2+m_2{}^2-m_1{}^2)^2-\tfrac{1}{3}P^2m_2{}^2\right.\right.$$

$$\left.-\frac{1}{4P^2}(P^2+m_2{}^2-m_1{}^2)([w(P^2, m_1{}^2, m_2{}^2)]^2-2P^2m_2{}^2)\right]$$

$$-\tfrac{1}{3}P^\mu g^{\nu\lambda}[w(P^2, m_1{}^2, m_2{}^2)]^2$$

$$\left.-(P^\mu g^{\nu\lambda}+P^\nu g^{\mu\lambda}+P^\lambda g^{\mu\nu})\frac{1}{24P^2}(P^2+m_2{}^2-m_1{}^2)[w(P^2, m_1{}^2, m_2{}^2)]^2\right\}$$

$$\int \frac{d^3p_1}{2E_1}\frac{d^3p_2}{2E_2}\,\delta^{(4)}(P-p_1-p_2)\,\delta[(p-p_1)^2-t]f(p_1p_2)$$

$$= \frac{\pi}{2}\frac{1}{w[P^2, p^2, (P-p)^2]}f[\tfrac{1}{2}(P^2-m_1{}^2-m_2{}^2)]$$

3. Three-Particle Integrals

Let $z=\sqrt{P^2}$

$$\int \frac{d^3p_1}{2E_1}\frac{d^3p_2}{2E_2}\frac{d^3p_3}{2E_3}\,\delta^{(4)}(P-p_1-p_2-p_3)f(p_1p_2)$$

$$= \frac{\pi^2}{4}\frac{1}{P^2}\int_{(m_1+m_2)^2}^{(z-m_3)^2}\frac{ds}{s}\,w(s, m_1{}^2, m_2{}^2)w(P^2, m_3{}^2, s)f[\tfrac{1}{2}(s-m_1{}^2-m_2{}^2)]$$

$$\int \frac{d^3p_1}{2E_1}\frac{d^3p_2}{2E_2}\frac{d^3p_3}{2E_3}\,\delta^{(4)}(P - p_1 - p_2 - p_3)f(p_1p_2)p_3{}^{\mu}$$

$$= \frac{\pi^2}{8}\frac{P^\mu}{P^4}\int\limits_{(m_1+m_2)^2}^{(z-m_3)^2}\frac{ds}{s}\,w(s, m_1{}^2, m_2{}^2)w(P^2, m_3{}^2, s)[P^2 + m_3{}^2 - s]$$

$$\times f[\tfrac{1}{2}(s - m_1{}^2 - m_2{}^2)]$$

$$\int \frac{d^3p_1}{2E_1}\frac{d^3p_2}{2E_2}\frac{d^3p_3}{2E_3}\,\delta^{(4)}(P - p_1 - p_2 - p_3)f(p_1p_2)p_1{}^{\mu}$$

$$= \frac{\pi^2}{16}\frac{P^\mu}{P^4}\int\limits_{(m_1+m_2)^2}^{(z-m_3)^2}\frac{ds}{s^2}\,w(s, m_1{}^2, m_2{}^2)w(P^2, m_3{}^2, s)(s + m_1{}^2 - m_2{}^2)$$

$$\times (P^2 + s - m_3{}^2)f[\tfrac{1}{2}(s - m_1{}^2 - m_2{}^2)]$$

$$\int \frac{d^3p_1}{2E_1}\frac{d^3p_2}{2E_2}\frac{d^3p_3}{2E_3}\,\delta^{(4)}(P - p_1 - p_2 - p_3)f(p_1p_2)(p_1p_3)$$

$$= \frac{\pi^2}{16}\frac{1}{P^2}\int\limits_{(m_1+m_2)^2}^{(z-m_3)^2}\frac{ds}{s^2}\,w(s, m_1{}^2, m_2{}^2)w(P^2, m_3{}^2, s)(s + m_1{}^2 - m_2{}^2)$$

$$\times (P^2 - m_3{}^2 - s)f[\tfrac{1}{2}(s - m_1{}^2 - m_2{}^2)]$$

$$\int \frac{d^3p_1}{2E_1}\frac{d^3p_2}{2E_2}\frac{d^3p_3}{2E_3}\,\delta^{(4)}(P - p_1 - p_2 - p_3)f(p_1p_2)(p_1p_3)^2$$

$$= \frac{\pi^2}{16}\frac{1}{P^2}\int\limits_{(m_1+m_2)^2}^{(z-m_3)^2}\frac{ds}{s^2}\,w(s, m_1{}^2, m_2{}^2)w(P^2, m_3{}^2, s)$$

$$\times \left\{\frac{1}{3s}[w(s, m_1{}^2, m_2{}^2)w(P^2, m_3{}^2, s)]^2 + m_1{}^2[w(P^2, m_3{}^2, s)]^2\right.$$

$$\left. + m_3{}^2[w(s, m_1{}^2, m_2{}^2)]^2 + 4sm_1{}^2m_3{}^2\right\} f[\tfrac{1}{2}(s - m_1{}^2 - m_2{}^2)]$$

$$P^2 - m_1{}^2 - m_2{}^2 - m_3{}^2 = 2(p_1p_2 + p_1p_3 + p_2p_3)$$

4. n-Particle Integrals

$$\int \frac{d^3p_1}{2E_1} \cdots \frac{d^3p_n}{2E_n} \delta^{(4)}(P - \Sigma p_i)$$

$$= \left(\frac{\pi}{2}\right)^{n-1} \int\limits_{(m_1+m_2)^2}^{z_1} \frac{ds_1}{s_1} w(s_1, m_1^2, m_2^2) \int\limits_{(\sqrt{s_1}+m_3)^2}^{z_2} \frac{ds_2}{s_2} w(s_1, s_2, m_3^2)$$

$$\times \int\limits_{(\sqrt{s_2}+m_4)^2}^{z_3} \frac{ds_3}{s_3} w(s_2, s_3, m_4^2) \cdots \int\limits_{(\sqrt{s_{n-3}}+m_{n-1})^2}^{z_{n-2}} \frac{ds_{n-2}}{s_{n-2}} w(s_{n-3}, s_{n-2}, m_{n-1}^2)$$

$$\times \frac{1}{P^2} w(s_{n-2}, P^2, m_n^2)$$

$$z_1 = (\sqrt{P^2} - m_3 - m_4 - \cdots - m_n)^2$$

$$z_2 = (\sqrt{P^2} - m_4 - m_5 - \cdots - m_n)^2$$

$$\vdots$$

$$z_{n-2} = (\sqrt{P^2} - m_n)^2$$

VII. Weak Interaction Hamiltonian and S-Operator

1. Standard V-A Theory

The current-current Hamiltonian (CC) is

$$\mathscr{H}_W(x) = \frac{G}{\sqrt{2}} J_\lambda^\dagger(x) J^\lambda(x)$$

$$J_\lambda(x) = l_\lambda(x) + \cos\vartheta\, j_\lambda^\pi(x) + \sin\vartheta\, j_\lambda^K(x)$$

ϑ is the Cabibbo angle

$$l_\lambda(x) = \sum_l \bar{\psi}_l(x)\gamma_\lambda(1+\gamma_5)\psi_{\nu_l}(x), \qquad l = e, \mu$$

The corresponding S-operator is

$$S = I - i\frac{G}{\sqrt{2}} \int d^4x\, J_\lambda^\dagger(x) J^\lambda(x) + 0(G^2)$$

The value of G is [2]

$$\begin{aligned} G &= 1.026 \cdot 10^{-5}\, M_p^{-2} \\ &= 1.166 \cdot 10^{-5}\, (\mathrm{GeV})^{-2} \\ &= 4.5403 \cdot 10^{-33}\, \mathrm{cm}^2 \end{aligned}$$

The value of the Cabibbo angle is [3]

$$\vartheta = 0.239 \pm 0.005$$

The intermediate vector boson Hamiltonian (IVB) is

$$\mathscr{H}_{\mathrm{IVB}}(x) = g[J_\lambda^\dagger(x) W^\lambda(x) + J^\lambda(x) W_\lambda^\dagger(x)]$$

In the limit of small invariant momentum transfer or large W-mass the relation becomes true

$$\frac{G}{\sqrt{2}} = \frac{g^2}{M_W^2}$$

2. Unified Models

Representative for all unified models we select the original Salam-Weinberg model whose leptonic Lagrangian in the physical U-gauge is

$$\mathcal{L}_{SW} = \mathcal{L}_0 + \sum_{l=e,\mu} \left\{ \left[\frac{g}{2\sqrt{2}} \bar{l}\gamma_\lambda(1+\gamma_5)\nu_l W^\lambda + \text{h.c.} \right] + e \cdot \bar{l}\gamma_\mu l A^\mu \right.$$

$$\left. + \frac{g}{4\cos\varphi} [\bar{\nu}_l\gamma_\lambda(1+\gamma_5)\nu_l - \bar{l}\gamma_\lambda(C_v+\gamma_5)l] Z^\lambda \right\}$$

$$- h\lambda^2\chi^2 - h\lambda\chi^3 - \frac{h}{4}\chi^4 + \tfrac{1}{2}(\partial_\mu\chi)^2$$

$$+ \frac{g^2}{4}\left(W_\mu{}^\dagger W^\mu + \frac{1}{2\cos^2\varphi} Z_\mu Z^\mu \right)(\lambda^2 + 2\lambda\chi + \chi^2)$$

$$+ \sum_{l=e,\mu} m_l \left(1 + \frac{1}{\lambda}\chi\right) \bar{l}l$$

$C_v = 1 - 4\sin^2\varphi$
$W_\lambda, W_\lambda{}^\dagger$ = charged intermediate boson fields
A_μ = photon field
Z_λ = neutral intermediate boson field
χ = scalar Higgs field

φ is the Salam-Weinberg angle.

$$\mathcal{L}_0 = \tfrac{1}{4}(\mathbf{A}_{\mu\nu})^2 + \tfrac{1}{4}(B_{\mu\nu})^2 + i\sum_{l=e,\mu} \bar{l}\gamma^\mu\partial_\mu l$$

$$\mathbf{A}_{\mu\nu} = \partial_\mu\mathbf{A}_\nu - \partial_\nu\mathbf{A}_\mu + g[\mathbf{A}_\mu \times \mathbf{A}_\nu]$$

$$B_{\mu\nu} = \partial_\mu B_\nu - \partial_\nu B_\mu$$

$$A_1{}^\mu + iA_2{}^\mu = \sqrt{2}\, W^\mu$$

$$A_3{}^\mu = \cos\varphi\, Z^\mu - \sin\varphi\, A^\mu$$

$$B^\mu = \sin\varphi\, Z^\mu + \cos\varphi\, A^\mu$$

The following relations hold:

$$e = g\sin\varphi = g'\cos\varphi$$

$$\frac{G}{\sqrt{2}} = \frac{1}{2\lambda^2} = \frac{g^2}{8M_W{}^2} = \frac{e^2}{8M_W{}^2\sin^2\varphi}$$

$$M_W = \frac{g\lambda}{2} = \frac{e\lambda}{2 \sin \varphi} > 37.3\,\text{GeV}$$

$$M_Z = \frac{M_W}{\cos \varphi} > 74.6\,\text{GeV}$$

$$m_\chi = \lambda\sqrt{2h}$$

VIII. Discrete Transformations

1. Space Reflections

$$\psi_P(x) = P\psi(x)P^{-1} = \gamma_0\psi(x_0, -\mathbf{x})$$

$$P^\dagger = P^{-1} = P$$

$$Pa_s(\mathbf{p})P^{-1} = a_s(-\mathbf{p})$$

$$Pb_s^\dagger(\mathbf{p})P^{-1} = -b_s^\dagger(-\mathbf{p})$$

for bosons with intrinsic parity η_P:

$$\phi_P(x) = P\phi(x)P^{-1} = \eta_P\phi(x_0, -\mathbf{x})$$

$$P\alpha(\mathbf{k})P^{-1} = \eta_P\alpha(\mathbf{k})$$

$$P\beta^\dagger(\mathbf{k})P^{-1} = \eta_P\beta^\dagger(\mathbf{k})$$

2. Charge Conjugation

$$\psi_C(x) = C\psi(x)C^{-1} = \mathscr{C}\bar{\psi}^T(x)$$

$$\bar{\psi}_C(x) = C\bar{\psi}(x)C^{-1} = \psi^T(x)\mathscr{C}^{-1}$$

$$C = C^\dagger = C^{-1}$$

$$Ca_s(\mathbf{p})C^{-1} = b_s(\mathbf{p})$$

$$Cb_s(\mathbf{p})C^{-1} = a_s(\mathbf{p})$$

$$\mathscr{C}^{-1}\gamma_\mu\mathscr{C} = -\gamma_\mu{}^T$$

$$\mathscr{C}^{-1}\gamma_5\mathscr{C} = \gamma_5{}^T$$

$$\bar{u}\mathscr{C}^T = v^T$$

$$\mathscr{C}^T = -\mathscr{C}$$

3. Time Reflection

The antilinear operator T is always understood to act on the ket-vector, i.e.

$$\langle\alpha|T|\beta\rangle = \langle\alpha|T\beta\rangle$$

if it is to act on the bra, the matrix element has to be complex conjugated.

$$\langle\alpha|T^\dagger|\beta\rangle = \langle T\alpha|\beta\rangle^*$$

furthermore, for any complex number λ,

$$T\lambda|\alpha\rangle = \lambda^* T|\alpha\rangle$$
$$T^\dagger = T^{-1}.$$

Invariance of the S-operator means

$$(TST^{-1})^\dagger = S$$
$$\psi_t(x) = [T\psi(x)T^{-1}]^\dagger = \bar{\psi}(-x_0, \mathbf{x})\gamma_0\gamma_5\mathscr{C}$$
$$\bar{\psi}_t(x) = [T\bar{\psi}(x)T^{-1}]^\dagger = \mathscr{C}^{-1}\gamma_5\gamma_0\psi(-x_0, \mathbf{x})$$
$$Ta_r(\mathbf{p})T^{-1} = a_s(-\mathbf{p})$$
$$Tb_r(\mathbf{p})T^{-1} = b_s(-\mathbf{p}) \qquad r \neq s$$

Let the S-operator be

$$S = I + iR$$

and let $|\alpha\rangle$ be an eigenstate of the (strong) S-operator

$$S|\alpha\rangle = \exp(2i\delta_\alpha)|\alpha\rangle$$

then the Fermi-Watson theorem states [4]

$$\langle T\alpha|R|T\beta\rangle^* = \exp(-2i\delta_\alpha)\,\langle\alpha|R|\beta\rangle$$

4. *G*-Parity

Let I_k be the components of the isospin-operator.

$$G = C\exp(i\pi I_2)$$

in particular for nucleons:

$$\exp\left(i\frac{\pi}{2}\tau_2\right) = i\tau_2$$
$$G|p\rangle = |\bar{n}\rangle$$
$$G|n\rangle = -|\bar{p}\rangle$$

for π-mesons of any charge:

$$G|\pi^i\rangle = -|\pi^i\rangle \qquad i = +, -, 0$$

IX. Muon-Decay

1. The Matrix Elements

$$\mu^- \to e^- + \bar{\nu}_e + \nu_\mu$$

$$\mu^+ \to e^+ + \nu_e + \bar{\nu}_\mu$$

$$16 m_e m_\mu m_\nu^2 \sum_{\text{Spins}} |\bar{u}_e(q_1)\gamma_\lambda(1+\gamma_5)v_{\nu_e}(q_2)\cdot \bar{u}_{\nu_\mu}(q_3)\gamma^\lambda(1+\gamma_5)u_\mu(p)|^2 = 256(pq_2)(q_1q_3)$$

where the neutrinos have been given a small mass for normalization purposes.

2. The Electron Spectrum

Let E be the electron total energy, q its momentum, and P the muon polarization.

$$x = \frac{E}{E_0}$$

$$E_0 = \frac{m_\mu^2 + m_e^2}{2m_\mu}$$

$$\varepsilon = \frac{m_e}{m_\mu}$$

$$N(E,\Omega) = \frac{G^2}{12\pi^4} m_\mu q E E_0 \left\{3 - 3x + \tfrac{2}{3}\rho(4x-3) + 6\eta\varepsilon\left(\frac{1}{x} - 1\right)\right.$$

$$\left. - P\frac{q}{E}\cos\vartheta\cdot\xi[1 - x + 2\delta(\tfrac{4}{3}x - 1)]\right\} + 0(\varepsilon^2)$$

with

$\rho = \frac{3}{4}$ (Michèl-parameter)

$\delta = \frac{3}{4}$ (shape-parameter)

$\xi = -1$ (asymmetry parameter)

$\eta = 0$ (eta-parameter)

The experiment of S. E. DERENZO et al. [5] gives

$$\rho = 0.7518 \pm 0.0026$$

$$\delta = 0.7540 \pm 0.0085$$

$$\xi = -\ 0.973 \pm 0.014$$

$$\eta = -\ 0.12 \pm 0.21$$

$$\Gamma(\mu) = \frac{1}{\tau_\mu} = \int_{m_e}^{E_0} dE \int d\Omega\, N(E, \Omega) = \frac{G^2 m_\mu^{\ 5}}{192\pi^3} + 0(\varepsilon)$$

3. Effects of the IVB

Let

$$\zeta = \frac{m_\mu^{\ 2}}{M_W^{\ 2}}$$

$$N(x, \cos\vartheta) = \frac{G^2 m_\mu^{\ 5}}{192\pi^3} x^2\{(3 - 2x) + \zeta x(2 - x)$$

$$+ P\cos\vartheta(1 - 2x - \zeta x^2)\} + 0(\varepsilon, \zeta^2)$$

$$G_\mu = G(1 + \tfrac{3}{10}\zeta) + 0(\varepsilon, \zeta^2)$$

Approximately valid is

$$\rho \doteq \tfrac{3}{4} + \tfrac{1}{3}\zeta$$

X. Semi-Leptonic Baryon Decay

1. List of Decay-Processes and Selection Rules

The hypercharge conserving semi-leptonic baryon decays are:

$$\left.\begin{aligned} n &\to p + e^- + \bar{\nu}_e \\ \Sigma^- &\to \Lambda + e^- + \bar{\nu}_e \\ \Sigma^+ &\to \Lambda + e^+ + \nu_e \end{aligned}\right\} \text{observed}$$

$$\Sigma^0 \to \Sigma^+ + e^- + \bar{\nu}_e$$

$$\Sigma^- \to \Sigma^0 + e^- + \bar{\nu}_e$$

$$\Xi^- \to \Xi^0 + e^- + \bar{\nu}_e$$

The hypercharge changing semi-leptonic decays of baryons with 3 final particles are (channels in brackets compete with electromagnetic decay channels!):

$$\Lambda \to p + l^- + \bar{\nu}_l \quad \text{observed}$$

$$(\Sigma^0 \to p + l^- + \bar{\nu}_l)$$

$$\Sigma^- \to n + l^- + \bar{\nu}_l \quad \text{observed}$$

$$\Xi^0 \to \Sigma^+ + l^- + \bar{\nu}_l$$

$$\Xi^- \to \Lambda + l^- + \bar{\nu}_l \quad \text{observed}$$

$$\Xi^- \to \Sigma^0 + l^- + \bar{\nu}_l$$

$$\Omega^- \to \Xi^0 + l^- + \bar{\nu}_l$$

The hypercharge changing semi-leptonic decays of baryons with more than 3 final particles are:

$$\Lambda \to p + \pi^0 + e^- + \bar{\nu}_e$$

$$\Lambda \to n + \pi^+ + e^- + \bar{\nu}_e$$

$$(\Sigma^0 \to p + \pi^0 + l^- + \bar{\nu}_l)$$

$$(\Sigma^0 \to n + \pi^+ + l^- + \bar{\nu}_l)$$

$$\Sigma^- \to n + \pi^0 + l^- + \bar{\nu}_l$$
$$\Sigma^- \to p + \pi^- + l^- + \bar{\nu}_l$$
$$\Xi^- \to \Lambda + \pi^0 + e^- + \bar{\nu}_e$$
$$\Omega^- \to \Xi^0 + \pi^0 + l^- + \bar{\nu}_l$$
$$\Omega^- \to \Xi^- + \pi^+ + l^- + \bar{\nu}_l$$
$$\Omega^- \to \Xi^0 + \pi^0 + \pi^0 + e^- + \bar{\nu}_e$$
$$\Omega^- \to \Xi^0 + \pi^+ + \pi^- + e^- + \bar{\nu}_e$$
$$\Omega^- \to \Xi^- + \pi^+ + \pi^0 + e^- + \bar{\nu}_e$$

The selection rules are

$$|\Delta Y| < 2$$
$$\Delta Y = \Delta Q_H \quad \text{for} \quad |\Delta Y| = 1$$

ΔQ_H is the change of charge of the hadrons!

$$\Delta I = |\mathbf{I}_f - \mathbf{I}_i| = \tfrac{1}{2} \quad \text{for} \quad |\Delta Y| = 1.$$

Forbidden by these selection rules are the following baryon decays:

$$\Sigma^+ \nrightarrow n + l^+ + \nu_l$$
$$\Xi^0 \nrightarrow \Sigma^- + l^+ + \nu_l$$
$$\Lambda \nrightarrow n + \pi^- + e^+ + \nu_e$$
$$\Sigma^0 \nrightarrow n + \pi^- + l^+ + \nu_l$$
$$\Sigma^+ \nrightarrow p + \pi^- + l^+ + \nu_l$$
$$\Sigma^+ \nrightarrow n + \pi^0 + l^+ + \nu_l$$
$$\Omega^- \nrightarrow \Xi^- + \pi^- + l^+ + \nu_l$$
$$\Omega^- \nrightarrow \Xi^- + \pi^- + \pi^0 + e^+ + \nu_e$$
$$\Omega^- \nrightarrow \Xi^0 + \pi^- + \pi^- + e^+ + \nu_e.$$

2. Matrix-Element and Decay Spectra

The matrix element of the hadronic weak current for the decay

$$A \to B + l^- + \bar{\nu}_l$$

is (with the definitions of Sec. III.4)

$$\langle B(p_2)|j_\lambda^\dagger(x)|A(p_1)\rangle = \frac{1}{(2\pi)^3}\left(\frac{m_A m_B}{E_A E_B}\right)^{1/2} \bar{u}(p_2)\Gamma_\lambda(p_1, p_2)u(p_1)\exp(iqx)$$

Let

$$p = p_2 + p_1$$

$$q = p_2 - p_1$$

$$\Gamma_\lambda(p_1, p_2) = F_1(q^2)\gamma_\lambda + F_2(q^2)i\sigma_{\lambda\nu}q^\nu + F_3(q^2)q_\lambda + G_1(q^2)\gamma_\lambda\gamma_5 + G_2(q^2)q_\lambda\gamma_5 + G_3(q^2)i\sigma_{\lambda\nu}q^\nu\gamma_5$$

With

$$\Sigma = m_A + m_B$$

$$\Delta = m_A - m_B$$

the following relations hold between the covariants:

$$\bar{u}(p_2)\not{p}u(p_1) = \Sigma\bar{u}(p_2)u(p_1)$$

$$\bar{u}(p_2)\not{q}u(p_1) = -\Delta\bar{u}(p_2)u(p_1)$$

$$\bar{u}(p_2)i\sigma_{\mu\nu}p^\mu q^\nu u(p_1) = (q^2 - \Delta^2)\bar{u}(p_2)u(p_1)$$

$$\bar{u}(p_2)i\gamma_\lambda\sigma_{\mu\nu}p^\mu p^\nu u(p_1) = (q^2 - \Sigma^2)\bar{u}(p_2)\gamma_\lambda u(p_1) + 2m_A\bar{u}(p_2)(p + q)_\lambda u(p_1)$$

General rule: Relations for axial quantities (with an extra γ_5) can be obtained by the replacement $\Delta \leftrightarrow -\Sigma$.

$$\bar{u}(p_2)[i\sigma_{\mu\nu}q^\nu + p_\mu - \gamma_\mu\Sigma]u(p_1) = 0$$

$$\bar{u}(p_2)[i\sigma_{\mu\nu}p^\nu + q_\mu + \gamma_\mu\Delta]u(p_1) = 0$$

$$\bar{u}(p_2)[i\sigma_{\mu\nu}q^\nu + p_\mu + \gamma_\mu\Delta]\gamma_5 u(p_1) = 0$$

$$\bar{u}(p_2)[i\sigma_{\mu\nu}p^\nu + q_\mu - \gamma_\mu\Sigma]\gamma_5 u(p_1) = 0$$

Def.:

$$a = \begin{cases}\cos\vartheta \\ \sin\vartheta\end{cases} \text{for} \quad |\Delta Y| = \begin{cases}0 \\ 1\end{cases}$$

$$s = q^2 = \Delta^2 - 2m_A T_B$$

$$T_B = E_B - m_B$$

The recoil spectrum is

$$N(s) = \frac{1}{384\pi^3} \frac{G^2}{m_A{}^3} a^2 \frac{(s - m_l{}^2)^2}{s^2} \sqrt{(\Sigma^2 - s)(\Delta^2 - s)}$$

$$\times \Big\{ |F_1(s)|^2 \Big[\Delta^2(4s - m_l{}^2) + 2\Sigma^2\Delta^2 \frac{s + 2m_l{}^2}{s} - (\Sigma^2 + 2s)(2s + m_l{}^2) \Big]$$

$$+ |F_2(s)|^2(\Delta^2 - s)(2\Sigma^2 + s)(2s + m_l{}^2)$$

$$+ |F_3(s)|^2 \cdot 3m_l{}^2 s(\Sigma^2 - s)$$

$$+ 2 \operatorname{Re} F_1(s)F_2{}^*(s) \cdot 3\Sigma(\Delta^2 - s)(2s + m_l{}^2)$$

$$- 2 \operatorname{Re} F_1(s)F_3{}^*(s) \cdot 3m_l{}^2\Delta(\Sigma^2 - s)$$

$$+ |G_1(s)|^2 \Big[\Sigma^2(4s - m_l{}^2) + 2\Sigma^2\Delta^2 \frac{s + 2m_l{}^2}{s} - (\Delta^2 + 2s)(2s + m_l{}^2) \Big]$$

$$+ |G_2(s)|^2 \cdot 3m_l{}^2 s(\Delta^2 - s)$$

$$+ |G_3(s)|^2(\Sigma^2 - s)(2\Delta^2 + s)(2s + m_l{}^2)$$

$$+ 2 \operatorname{Re} G_1(s)G_2{}^*(s) \cdot 3m_l{}^2\Sigma(\Delta^2 - s)$$

$$- 2 \operatorname{Re} G_1(s)G_3{}^*(s) \cdot 3\Delta(2s + m_l{}^2)(\Sigma^2 - s) \Big\}$$

$$\Gamma(A; B, l) = \frac{1}{\tau_{A,Bl}} = \int_{m_l^2}^{\Delta^2} ds\, N(s)$$

$$= \frac{G^2 a^2 \Delta^5}{60\pi^3} \left(\frac{\Sigma}{2m_A}\right)^3 |F_1(0)|^2 \left[1 + 3 \left| \frac{G_1(0)}{F_1(0)} \right|^2 \right] r(x) + 0(\delta^2)$$

$$\delta = \frac{\Delta}{\Sigma}$$

$$r(x) = \frac{\Gamma(A; B, \mu)}{\Gamma(A; B, e)} = \tfrac{1}{2}(1 - x^2)^{3/2}(2 - 7x^2)$$

$$+ \frac{15}{4} x^4 \left[\log \frac{1 + \sqrt{1 - x^2}}{1 - \sqrt{1 - x^2}} - 2\sqrt{1 - x^2} \right]$$

$$x = \frac{m_l}{\Delta}$$

Values for the parameters are

A, B	δ	x	r	$r_{\exp}$	Ref.
Λ, p	0.086	0.6	0.16	0.19 ± 0.04	[7]
Σ^-, n	0.121	0.41	0.46	0.40 ± 0.04	[7]
Ξ^-, Λ	0.084	0.51	0.27	—	

3. Restrictions on the Form Factors

Time reversal invariance requires

$$[\mathscr{C}^{-1}\gamma_5\Gamma_\lambda(\eta p_1, \eta p_2)\gamma_5\mathscr{C}]^* = \eta_{(\lambda)}\Gamma_\lambda(p_1, p_2)$$

$$\eta_{(\lambda)} = \begin{cases} +1 \\ -1 \end{cases} \text{for } \lambda = \begin{cases} 0 \\ 1, 2, 3 \end{cases}$$

Consequently, all form factors are real.

Define a "first class current" by

$$Gj_\lambda{}^i G^{-1} = \varepsilon_i j_\lambda{}^i$$

$$\varepsilon_i = \begin{cases} +1 \\ -1 \end{cases} \text{for } i = \begin{cases} V \\ A \end{cases}$$

Currents with opposite sign in ε_i are "second class currents." If the current operators are of first class, the matrix element has to obey Weinberg's equation [6]:

$$\Gamma_\lambda{}^i(p_1, p_2) = -\varepsilon_i\mathscr{C}\Gamma_\lambda{}'^{iT}(-p_2, -p_1)\mathscr{C}^{-1}$$

where the prime on the vertex function of the right-hand side denotes the G-parity related process.

Distinguish

TWIM: transitions within iso-multiplets ($\Delta = 0$)

and

TBIM: transitions between iso-multiplets with equal hypercharge ($\Delta \neq 0$).

Examples: TWIM: $p \to n$

TBIM: $\Sigma \to \Lambda$

For TWIM, $F_3 = G_3 = 0$ for first class currents.
For TBIM, e.g.:

$$\Gamma_\lambda{}^i(\Sigma^+\Lambda) = -\varepsilon_i\mathscr{C}\Gamma_\lambda{}^{iT}(\Sigma^-\Lambda)\mathscr{C}^{-1}$$

$$\frac{\Gamma(\Sigma^+\Lambda)}{\Gamma(\Sigma^-\Lambda)} = \left(\frac{\Delta^+}{\Delta^-}\right)^5 = 0.60$$

The Conserved Vector Current (CVC) requires

$$\Delta F_1(q^2) = q^2F_3(q^2)$$

TWIM: $F_3(q^2) = 0$

TBIM: $F_1(0) = 0$

4. The Cabibbo Current

The following assumptions are made:

(i) The total hadronic current j_λ transforms as a component of an octet of SU(3).

(ii) The vector part of j_λ is in the same octet as the electromagnetic current.

iii) The current is of the form

$$J_\lambda(x) = l_\lambda(x) + \cos\vartheta\, j_\lambda{}^{\pi}(x) + \sin\vartheta\, j_\lambda{}^{K}(x)$$

ϑ is the Cabibbo angle

Consequently:

$$F_1(0)_{AB} = C_f{}^{AB}$$

$$G_1(0)_{AB} = C_f{}^{AB} G_1{}^f + C_d{}^{AB} G_1{}^d$$

AB	$C_f{}^{AB}$	$C_d{}^{AB}$	$G_1(0)/F_1(0)$
np	1	1	$G_1{}^f + G_1{}^d$
$\Sigma\Lambda$	0	$\sqrt{\frac{2}{3}}$	pure A
Λp	$-\sqrt{\frac{3}{2}}$	$-(1/\sqrt{6})$	$G_1{}^f + \frac{1}{3}G_1{}^d$
$\Sigma^- n$	-1	1	$G_1{}^f - G_1{}^d$
$\Sigma^0 p$	$-(1/\sqrt{2})$	$1/\sqrt{2}$	$G_1{}^f - G_1{}^d$
$\Xi^-\Lambda$	$\sqrt{\frac{3}{2}}$	$-(1/\sqrt{6})$	$G_1{}^f - \frac{1}{3}G_1{}^d$

Def.:

$$f_i = F_i(0) \cdot \Sigma / F_1(0) \qquad i = 2, 3$$

$$g_i = G_i(0) \cdot \Sigma / F_1(0)$$

AB	f_2
np	$\mu_p - \mu_n$
Λp	μ_p
$\Sigma^- n$	$\mu_p + 2\mu_n$
$\Xi^-\Lambda$	$\mu_p - \mu_n$

where μ_p and μ_n are proton and neutron anomalous magnetic moments.

Experimental values are

AB	$G_1(0)/F_1(0)$	Ref.
np	1.251 ± 0.009	14
Λp	0.66 ± 0.06	7
$\Sigma^- n$	$\|0.435 \pm 0.035\|$	15

5. Polarization Effects

The decay spectrum of baryons with polarization P is:

Def.:

$$\lambda = \frac{G_1(0)}{F_1(0)}$$

$$N(E, \vartheta, \varphi_l, \varphi_\nu) = \frac{G^2 a^2}{(2\pi)^5} (E_0 - E)^2 p^2 |F_1(0)|^2 (1 + 3\lambda^2 - 4\lambda g_3 \delta)$$

$$\times \{1 + a_{l\nu} \cos\vartheta + P[\alpha_l \cos\varphi_l + \alpha_\nu \cos\varphi_\nu]\}$$

where E and p are lepton energy and momentum.

$$E_0 = \frac{\Sigma\Delta + m_l{}^2}{2M_A}$$

$$a_{l\nu} = \frac{1 - \lambda^2 - 2\delta[1 + 3\lambda^2 - 2\lambda g_3]}{1 + 3\lambda^2 - 4\lambda g_3 \delta} + 0(\delta^2)$$

$$\alpha_l = \frac{2\lambda(1 - \lambda) - \frac{2}{3}\delta[1 + \lambda^2 + 2\lambda + f_2(1 + \lambda) + g_3(1 - 5\lambda)]}{1 + 3\lambda^2 - 4\lambda g_3 \delta} + 0(\delta^2)$$

$$\alpha_\nu = \frac{2\lambda(1 + \lambda) + \frac{2}{3}\delta[1 + \lambda^2 - 2\lambda + f_2(1 - \lambda) - g_3(1 + 5\lambda)]}{1 + 3\lambda^2 - 4\lambda g_3 \delta} + 0(\delta^2)$$

$$\alpha_B = \frac{-\frac{5}{2}\lambda + \frac{5}{3}\delta[\lambda(1 + f_2) + \frac{1}{2}g_3]}{1 + 3\lambda^2 - 4\lambda g_3 \delta} + 0(\delta^2)$$

The asymmetry parameters are connected to up-down asymmetries by

$$\alpha_i = 2\frac{N_+ - N_-}{N_+ + N_-} \qquad i = l, \nu, B$$

N_+ and N_- are the numbers of decays with the relevant particle (i) emitted into the half-sphere whose axis is parallel or anti-parallel to the spin of the decaying baryon.

The following sum-rule holds:

$$(1 - a_{l\nu}) - (\alpha_\nu - \alpha_l) = \frac{2}{3}\delta \frac{1 + 7\lambda^2 - 2f_2 - 2\lambda g_3}{1 + 3\lambda^2 - 4\lambda g_3 \delta} + 0(\delta^2)$$

XI. Semi-Leptonic Meson Decays

1. List of Decay Processes and Selection Rules

A decay process is called P_{ln} if the meson P decays into n particles and the lepton is of nature l.

The semi-leptonic pion decays are:

$$\pi^+ \to l^+ + \nu_l \qquad \pi^- \to l^- + \bar{\nu}_l$$

$$\pi^+ \to \pi^0 + e^+ + \nu_e \qquad \pi^- \to \pi^0 + e^- + \bar{\nu}_e$$

The semi-leptonic Kaon decays are:

$$K^+ \to l^+ + \nu_l \qquad K^- \to l^- + \bar{\nu}_l$$

$$K^+ \to \pi^0 + l^+ + \nu_l \qquad K^- \to \pi^0 + l^- + \bar{\nu}_l$$

$$K^+ \to \pi^0 + \pi^0 + l^+ + \nu_l \qquad K^- \to \pi^0 + \pi^0 + l^- + \bar{\nu}_l$$

$$K^+ \to \pi^+ + \pi^- + l^+ + \nu_l \qquad K^- \to \pi^+ + \pi^- + l^- + \bar{\nu}_l$$

$$K^+ \to 3\pi^0 + e^+ + \nu_e \qquad K^- \to 3\pi^0 + e^- + \bar{\nu}_e$$

$$K^+ \to \pi^+ + \pi^- + \pi^0 + e^+ + \nu_e \qquad K^- \to \pi^+ + \pi^- + \pi^0 + e^- + \bar{\nu}_e$$

$$K_{S,L} \to \pi^+ + l^- + \bar{\nu}_l$$

$$K_{S,L} \to \pi^- + l^+ + \nu_l$$

$$K_{S,L} \to \pi^+ + \pi^0 + l^- + \bar{\nu}_l$$

$$K_{S,L} \to \pi^- + \pi^0 + l^+ + \nu_l$$

$$K_{S,L} \to \pi^+ + 2\pi^0 + e^- + \bar{\nu}_e$$

$$K_{S,L} \to \pi^- + 2\pi^0 + e^+ + \nu_e$$

$$K_{S,L} \to \pi^+ + \pi^+ + \pi^- + e^- + \bar{\nu}_e$$

$$K_{S,L} \to \pi^- + \pi^- + \pi^+ + e^+ + \nu_e$$

Selection rules are

$$|\Delta Y| < 2$$

$$\Delta Y = \Delta Q_H \quad \text{for} \quad |\Delta Y| = 1$$

ΔQ_H is the change of charge of the hadrons!

$$\Delta I = |\mathbf{I}_f - \mathbf{I}_i| = \tfrac{1}{2} \quad \text{for} \quad |\Delta Y| = 1$$

Forbidden by these selection rules are the following meson decays:

$$K^+ \not\to \pi^+ + \pi^+ + l^- + \bar{\nu}_l$$

$$K^+ \not\to \pi^+ + \pi^+ + \pi^0 + e^- + \bar{\nu}_e$$

2. P_{l2}-Decays

The matrix element of the relevant hadronic current is

$$\langle 0|j_\lambda(x)|P(k)\rangle = \frac{i}{(2\pi)^{3/2}} \frac{1}{\sqrt{2k^0}} f_P k_\lambda \exp(-ikx)$$

The decay rate is

$$\Gamma(P_{l2}) = \frac{(Gaf_P)^2}{8\pi} \left(\frac{m_l}{m_P}\right)^2 \frac{(m_P{}^2 - m_l{}^2)^2}{m_P}$$

$$a = \begin{cases} \cos\vartheta \\ \sin\vartheta \end{cases} \quad \text{for} \quad |\Delta Y| = \begin{cases} 0 \\ 1 \end{cases}$$

Experimental values are [13]

$$f_\pi = (130.15 \pm 0.06) \text{ MeV}$$

$$= 0.932 \cdot m_\pi$$

The e-μ ratio is

$$R_P = \frac{\Gamma(P_{e2})}{\Gamma(P_{\mu 2})} = \left(\frac{m_e}{m_\mu}\right)^2 \left(\frac{m_P{}^2 - m_e{}^2}{m_P{}^2 - m_\mu{}^2}\right)^2$$

P	R_P	$(R_P)_{\exp}$	Ref.
π	$1.283 \cdot 10^{-4}$	$(1.247 \pm 0.028) \cdot 10^{-4}$	[8]
K	$2.1 \cdot 10^{-5}$	$\left(1.9 \begin{smallmatrix} +0.5 \\ -0.4 \end{smallmatrix}\right) \cdot 10^{-5}$	[9]

3. P_{l3}-Decays

The matrix element is

$$\langle \pi^0(k')|j_\lambda(x)|P(k)\rangle$$

$$= \frac{1}{(2\pi)^3}\frac{1}{2\sqrt{E_\pi E_P}}\{(k+k')_\lambda f_+(q^2) + (k-k')_\lambda f_-(q^2)\}\exp[ix(k'-k)]$$

$$q^2 = (k-k')^2$$

SU_3-symmetric values are (cf. Sec. X.4):

$$f_+^{\pi^+\pi^0}(0) = \sqrt{2} \qquad f_-^{\pi^+\pi^0}(q^2) = 0$$

$$f_+^{K^0\pi^\pm}(0) = 1$$

$$f_+^{K^\pm\pi^0}(0) = \frac{1}{\sqrt{2}}$$

Parametrization of the form factors:

$$f_\pm(q^2) = f_\pm(0)\left\{1 + \lambda_\pm \frac{q^2}{m_\pi^2} + \cdots\right\}$$

The scalar form factor is defined:

$$f_0(q^2) = f_+(q^2) + \frac{q^2}{m_K^2 - m_\pi^2} f_-(q^2)$$

$$f_0(q^2) = f_0(0)\left\{1 + \lambda_0 \frac{q^2}{m_\pi^2} + \cdots\right\}$$

$$\xi = \frac{f_-(0)}{f_+(0)}$$

The recoil energy spectrum for K_{l3} decay is (ω is the total π-energy, p_π its momentum)

$$N(\omega) = \frac{G^2 \sin^2\vartheta}{12\pi^3} m_K |f_+(s)|^2 p_\pi^3$$

$$s = m_K^2 + m_\pi^2 - 2m_K\omega$$

Let $\lambda_+ = \lambda_- = 0$:

$$\frac{\Gamma(K_{\mu 3})}{\Gamma(K_{e3})} = 0.646 + 0.1264\xi + 0.0192\xi^2 \geqslant 0.44$$

the lepton recoil spectrum:

$$n(\varepsilon) = \frac{G^2 \sin^2\vartheta}{32\pi^3} |f_+|^2 m_K{}^4 p_l \left(1 - \frac{m_\pi{}^2}{t}\right)^2$$

$$\times \{4\varepsilon(1 - 2\varepsilon) + 5\alpha^2\varepsilon - \alpha^4 + \xi\alpha^2(4 - 6\varepsilon + 2\alpha^2) + \xi^2\alpha^2(\varepsilon - \alpha^2)\}$$

with

$$t = m_K{}^2 + m_l{}^2 - 2m_K E_l$$

$$\alpha = \frac{m_l}{m_K}$$

$$\varepsilon = \frac{E_l}{m_K}$$

the longitudinal lepton polarization:

$$P_l(\varepsilon) = \frac{p_l}{m_K} \frac{N(\varepsilon, \alpha, \xi)}{D(\varepsilon, \alpha, \xi)}$$

$$N(\varepsilon, \alpha, \xi) = -4(1 - 2\varepsilon) - 3\alpha^2 + 2\alpha^2\xi + \alpha^2\xi^2$$

$$D(\varepsilon, \alpha, \xi) = 4\varepsilon(1 - 2\varepsilon) + 5\alpha^2\varepsilon - \alpha^4 + \xi\alpha^2(4 - 6\varepsilon + 2\alpha^2) + \xi^2\alpha^2(\varepsilon - \alpha^2)$$

XII. Current-Algebra

1. General Relation

Let

$$Q_1 = \int d^3x \, \psi^\dagger(x) \Lambda_1 \psi(x)$$

$$Q_2 = \int d^3x \, \psi^\dagger(x) \Lambda_2 \psi(x)$$

then the equal-time commutator is

$$[Q_1, Q_2] = \int d^3x \, \psi^\dagger(x) [\Lambda_1, \Lambda_2] \psi(x)$$

2. The Basic Commutators

Define

$$I^i = \int d^3x \, j_0^{V,i}(x) = \int d^3x \, \psi_q^\dagger(x) \tau^i \psi_q(x)$$

$$Q_a^i = \int d^3x \, j_0^{A,i}(x) = \int d^3x \, \psi_q^\dagger(x) \gamma_5 \tau^i \psi_q(x)$$

$$\psi_q = \text{quark spinor}$$

The algebra of SU(2) ⊗ SU(2) is

$$[I^+, I^-] = 2I^0$$

$$[I^0, I^\pm] = \pm I^\pm$$

$$[I^0, Q_a^\pm] = \pm Q_a^\pm$$

$$[Q_a^+, Q_a^-] = 2I^0$$

Equivalently, define

$$J_{\lambda\pm}{}^i(x) = \tfrac{1}{4}\bar{\psi}_q(x)\gamma_\lambda(1 \pm \gamma_5)\tau^i\psi_q(x)$$

then the algebra SU(2) ⊗ SU(2) is

$$[J_{0\pm}{}^l(x), J_{0\pm}{}^k(y)]_{x_0=y_0} = i\varepsilon^{lkm} J_{0\pm}{}^m(x)\,\delta^{(3)}(\mathbf{x}-\mathbf{y})$$

$$[J_{0\pm}{}^l(x), J_{0\mp}{}^k(y)]_{x_0=y_0} = 0$$

The algebra of SU(3) ⊗ SU(3) is

$$[J_{0\pm}{}^a(x), J_{0\pm}{}^b(y)]_{x_0=y_0} = if_{abc}J_{0\pm}{}^c(x)\,\delta^{(3)}(\mathbf{x}-\mathbf{y})$$

$$[J_{0\pm}{}^a(x), J_{0\mp}{}^b(y)]_{x_0=y_0} = 0$$

where

$$J_{\lambda\pm}{}^a(x) = \tfrac{1}{4}\bar{\psi}_q(x)\gamma_\lambda(1 \pm \gamma_5)\lambda^a\psi_q(x)$$

λ^a are the 8 Gell-Mann 3 × 3 matrices with the following properties:

$$[\lambda^a, \lambda^b] = 2if_{abc}\lambda^c$$

$$\{\lambda^a, \lambda^b\} = \tfrac{4}{3}\delta_{ab} + 2d_{abc}\lambda^c$$

$$\mathrm{Tr}\,\lambda^a[\lambda^b, \lambda^c] = 4if_{abc}$$

$$\mathrm{Tr}\,\lambda^a\{\lambda^b, \lambda^c\} = 4d_{abc}$$

The non-vanishing elements of f_{abc} are:

a	b	c	f_{abc}	a	b	c	f_{abc}
1	2	3	1	3	4	5	$\frac{1}{2}$
1	4	7	$\frac{1}{2}$	3	6	7	$-\frac{1}{2}$
1	5	6	$-\frac{1}{2}$	4	5	8	$\sqrt{3}/2$
2	4	6	$\frac{1}{2}$	6	7	8	$\sqrt{3}/2$
2	5	7	$\frac{1}{2}$				

The non-vanishing elements of d_{abc} are:

a	b	c	d_{abc}	a	b	c	d_{abc}
1	1	8	$1/\sqrt{3}$	3	5	5	$\frac{1}{2}$
1	4	6	$\frac{1}{2}$	3	6	6	$-\frac{1}{2}$
1	5	7	$\frac{1}{2}$	3	7	7	$-\frac{1}{2}$
2	2	8	$1/\sqrt{3}$	4	4	8	$-1/(2\sqrt{3})$
2	4	7	$-\frac{1}{2}$	5	5	8	$-1/(2\sqrt{3})$
2	5	6	$\frac{1}{2}$	6	6	8	$-1/(2\sqrt{3})$
3	3	8	$1/\sqrt{3}$	7	7	8	$-1/(2\sqrt{3})$
3	4	4	$\frac{1}{2}$	8	8	8	$-1/\sqrt{3}$

Quantum numbers of the quark states are:

	I	I_3	Y	S	B	Q
q_1	$\frac{1}{2}$	$\frac{1}{2}$	$\frac{1}{3}$	0	$\frac{1}{3}$	$\frac{2}{3}$
q_2	$\frac{1}{2}$	$-\frac{1}{2}$	$\frac{1}{3}$	0	$\frac{1}{3}$	$-\frac{1}{3}$
q_3	0	0	$-\frac{2}{3}$	-1	$\frac{1}{3}$	$-\frac{1}{3}$

3. The Goldberger-Treiman Relation

Let $\phi_\pi(x)$ be the charged π-meson field.

$$(\Box + m_\pi^2)\phi_\pi(x) = \sqrt{2}\, g_{\pi NN} j_\pi(x)$$

$$\langle p(p_2)|j_\pi(x)|n(p_1)\rangle = \frac{1}{(2\pi)^3}\frac{m_N}{\sqrt{E_p E_n}}\bar{u}(p_2)\gamma_5 K(q^2) u(p_1)\exp(iqx)$$

$$K(m_\pi^2) = 1$$

then

$$f_\pi = \frac{\sqrt{2}\, m_N \lambda}{g_{\pi NN} K(0)}$$

$$\lambda = \frac{G_1(0)}{F_1(0)}$$

The Goldberger-Treiman relation results in the approximation $K(0) \doteq 1$:

$$f_\pi \doteq \frac{\sqrt{2}\, m_N \lambda}{g_{\pi NN}}$$

4. The Adler-Weisberger Relation

The PCAC relation is

$$\partial^\mu j_\mu{}^A = m_\pi{}^2 f_\pi \phi_\pi$$

or, in terms of isospin components,

$$\partial^\mu j_\mu{}^{A,i} = m_\pi{}^2 \frac{f_\pi}{\sqrt{2}} \phi_\pi{}^i$$

consequently

$$\langle \alpha | Q_a | \beta \rangle = \frac{i m_\pi{}^2 f_\pi}{E_\beta - E_\alpha} \int d^3x \langle \alpha | \phi_\pi(x) | \beta \rangle$$

the Adler-Weisberger relation is

$$\frac{1}{\lambda^2} = 1 - \frac{4 m_N{}^2}{\pi g^2_{\pi NN} K^2(0)} \int\limits_{m_N + m_\pi}^{\infty} dW \frac{W}{W^2 - M_N{}^2}$$
$$\times \{\sigma_{\rm tot}^{\pi^+ p}(W, m_\pi{}^2 = 0) - \sigma_{\rm tot}^{\pi^- p}(W, m_\pi{}^2 = 0)\}$$

with the definition

$$\lambda = \frac{G_1(0)}{F_1(0)}$$

5. The Mathur-Okubo-Pandit-Callan-Treiman Relation

$$f_+{}^{K^\pm \pi^0}(m_K{}^2) + f_-{}^{K^\pm \pi^0}(m_K{}^2) = \frac{1}{\sqrt{2}} \frac{f_K}{f_\pi}$$

Let

$$f_0(q^2) = f_+(q^2) + \frac{q^2}{m_K{}^2 - m_\pi{}^2} f_-(q^2)$$

then

$$f_0{}^{K^\pm \pi^0}(m_K{}^2) = \frac{1}{\sqrt{2}} \frac{f_K}{f_\pi} - \left(\frac{m_\pi}{m_K}\right)^2 f_-{}^{K^\pm \pi^0}(m_K{}^2) + 0\left(\frac{m_\pi{}^4}{m_K{}^4}\right)$$

XIII. Hadronic Hyperon Decay

1. Matrix Element and Decay Rate

Define

$$\langle B'(p_2), \pi(k)|T|B(p_1)\rangle = -2\sqrt{m_B m_{B'}}\,\bar{u}(p_2)[A + B\gamma_5]u(p_1)$$

$$= -2\sqrt{m_B E_{B'}}\,\eta_{B'}{}^T\left[a_S + \frac{\boldsymbol{\sigma}\mathbf{p}_2}{|\mathbf{p}_2|}\,a_P\right]\eta_B$$

$$a_S = \left[\frac{E_{B'} + m_{B'}}{2E_{B'}}\right]^{1/2} A$$

$$a_P = \frac{|\mathbf{p}_2|}{[2E_{B'}(E_{B'} + m_{B'})]^{1/2}}\,B$$

With the definitions

$$\delta = \frac{\Delta}{\Sigma} = \frac{m_B - m_{B'}}{m_B + m_{B'}}$$

$$y = \frac{m_\pi}{\Delta}$$

the decay rate is

$$\Gamma_{B,B'\pi} = \frac{1}{2\pi}\left(\frac{\Sigma}{2m_B}\right)^3 \Delta\sqrt{1-y^2}\,\sqrt{1-\delta^2 y^2}\{|A|^2(1-\delta^2 y^2) + |B|^2\,\delta^2(1-y^2)\}$$

$$= \frac{1}{2\pi}\left(\frac{\Sigma}{2m_B}\right)^3 \Delta\sqrt{1-y^2}\,\sqrt{1-\delta^2 y^2}(1+\delta^2 - 2\delta^2 y^2)[|a_S|^2 + |a_P|^2]$$

$$= \frac{|\mathbf{k}|}{2\pi}\frac{E_{B'}}{m_B}[|a_S|^2 + |a_P|^2]$$

Values for the parameters are

BB'	δ	y	
$\Sigma^- n$	0.121	0.54	$\pi^\pm$-decays
$\Sigma^+ n$	0.117	0.56	
Λp	0.086	0.79	
$\Xi^- \Lambda$	0.084	0.68	
$\Sigma^+ p$	0.118	0.54	π^0-decays
Λn	0.086	0.77	
$\Xi^0 \Lambda$	0.082	0.68	

2. Asymmetry Parameters

The decay spectrum of baryons with polarization P is:

$$N(\vartheta) \propto 1 + \alpha P \cos \vartheta$$

$$\langle P_{B'} \rangle = \text{polarization of decay baryon}$$

$$\langle P_{B'} \rangle = \frac{1}{1 + \alpha \mathbf{P} \hat{p}_2} \{(\alpha + \mathbf{P} \cdot \hat{p}_2)\hat{p}_2 + \beta(\mathbf{P} \times \hat{p}_2) + \gamma[\hat{p}_2 \times (\mathbf{P} \times \hat{p}_2)]\}$$

$$\hat{p}_2 = \frac{\mathbf{p}_2}{|\mathbf{p}_2|}$$

$$\alpha = \frac{2 \operatorname{Re} a_S^* a_P}{|a_S|^2 + |a_P|^2}$$

$$\beta = \frac{2 \operatorname{Im} a_S^* a_P}{|a_S|^2 + |a_P|^2}$$

$$\gamma = \frac{|a_S|^2 - |a_P|^2}{|a_S|^2 + |a_P|^2}$$

$$\alpha^2 + \beta^2 + \gamma^2 = 1$$

3. The $\Delta I = \frac{1}{2}$ Rule

$$2\tau(\Lambda, p\pi^-) = \tau(\Lambda, n\pi^0)$$

$$2\tau(\Xi^0, \Sigma^+\pi^-) = \tau(\Xi^-, \Sigma^0\pi^-)$$

$$a_{S,P}(\Sigma^+ \to n\pi^+) = a_{S,P}(\Sigma^- \to n\pi^-) + \sqrt{2}\, a_{S,P}(\Sigma^+ \to p\pi^0)$$

Experimental values are [7]

$$\left.\frac{\tau(\Lambda, n\pi^0)}{\tau(\Lambda, p\pi^-)}\right|_{\text{exp}} = 1.79 \pm 0.03$$

$$\left.\frac{\tau(\Xi^-, \Sigma^0\pi^-)}{\tau(\Xi^0, \Sigma^+\pi^-)}\right|_{\text{exp}} = 1.78 \pm 0.05$$

XIV. Quasi-Elastic Neutrino Scattering

1. Neutrino-Lepton Scattering

Cross-sections in the laboratory system with CC Hamiltonian are (section VII. 1)

$$\sigma(l\nu_l \to l\nu_l) = \frac{4G^2}{\pi}\frac{m_l E_\nu^2}{m_l + 2E_\nu} = \frac{2G^2 m_l}{\pi} E_\nu + 0\left(\frac{m_l}{E_\nu}\right)$$

$$\sigma(e\nu_e \to e\nu_e) \doteq 1.68 \cdot 10^{-41}\, E_\nu(\text{GeV})\ \text{cm}^2$$

$$\sigma(l\bar{\nu}_l \to l\bar{\nu}_l) = \frac{4G^2}{3\pi}\frac{m_l E_{\bar{\nu}}^2}{(m_l + 2E_{\bar{\nu}})^3}(4E_{\bar{\nu}}^2 + 6m_l E_{\bar{\nu}} + 3m_l^2) = \frac{2G^2 m_l}{3\pi} E_{\bar{\nu}} + 0\left(\frac{m_l}{E_{\bar{\nu}}}\right)$$

$$\sigma(e\bar{\nu}_e \to e\bar{\nu}_e) \doteq 0.56 \cdot 10^{-41}\, E_{\bar{\nu}}(\text{GeV})\ \text{cm}^2$$

$$\sigma(e\nu_\mu \to \mu\nu_e) = \frac{G^2}{\pi}\frac{(s - m_\mu^2)^2}{s} \qquad s = m_e(m_e + 2E_\nu)$$

$$\sigma(e\bar{\nu}_\mu \to e\bar{\nu}_\mu) = 0(G^4)$$

Let $T_l = E_l - m_l$ be the kinetic recoil energy of the lepton in the laboratory frame.

$$\frac{d\sigma(e\bar{\nu}_e \to e\bar{\nu}_e)}{dT_e} = \frac{2G^2 m_e}{\pi}\left(1 - \frac{T_e}{E_{\bar{\nu}}}\right)^2$$

The cross-section for $l\nu_l \to l\nu_l$ is pure $J = 0$.

Effects of the IVB on the center-of-momentum cross-sections are

$$\sigma(l\nu_l \to l\nu_l) = \frac{8g^4}{\pi}\frac{E^2}{M_W^2(M_W^2 + 4E^2)} + 0\left(\frac{m_l^2}{M_W^2}\right) \qquad CM,\ \text{IVB!}$$

$$\frac{d\sigma}{d\Omega}(l\nu_l \to l\nu_l) = \frac{2g^4}{\pi^2}\frac{E^2}{[2E^2(1 + \cos\vartheta) + M_W^2]^2} + 0\left(\frac{m_l^2}{M_W^2}\right) \qquad CM,\ \text{IVB!}$$

Matrix elements in the Salam-Weinberg theory are

$$M(e\bar{\nu}_e \to e\bar{\nu}_e) = \frac{G}{2\sqrt{2}} \bar{\nu}_e \gamma_\lambda (1 + \gamma_5) \nu_e \cdot \bar{e} \gamma^\lambda (C_V{}^{e\bar{\nu}_e} + \gamma_5) e$$

$$C_V{}^{e\bar{\nu}_e} = 1 + 4 \sin^2 \varphi$$

$$M(e\bar{\nu}_\mu \to e\bar{\nu}_\mu) = \frac{-G}{2\sqrt{2}} \bar{\nu}_\mu \gamma_\lambda (1 + \gamma_5) \nu_\mu \cdot \bar{e} \gamma^\lambda (C_V{}^{e\bar{\nu}_\mu} + \gamma_5) e$$

$$C_V{}^{e\bar{\nu}_\mu} = 1 - 4 \sin^2 \varphi$$

The recoil-spectrum in the Salam-Weinberg theory is

$$\frac{d\sigma(e\bar{\nu}_l \to e\bar{\nu}_l)}{dT_e} = \frac{G^2 m_e}{8\pi} \left[(C_V{}^{e\bar{\nu}_l} - 1)^2 + (C_V{}^{e\bar{\nu}_l} + 1)^2 \left(1 - \frac{T_e}{E_{\bar{\nu}}}\right)^2 \right.$$

$$\left. - ((C_V{}^{e\bar{\nu}_l})^2 - 1) \frac{m_e T_e}{E_{\bar{\nu}}{}^2} \right]$$

2. Neutrino-Baryon Scattering

Consider the process

$$\nu_l(p_2) + B(p_1) \to l(p_2') + B'(p_1')$$

The Mandelstam variables are

$$s + t + u = 2M_B{}^2 + m_l{}^2$$

$$s = M_B{}^2 + 2(p_1 p_2)$$

$$t = m_l{}^2 - 2(p_2 p_2')$$

$$u = m_l{}^2 + M_B{}^2 - 2(p_1 p_2')$$

The laboratory cross-section is

$$\frac{d\sigma}{dt} = \frac{G^2a^2}{16\pi}\frac{1}{M_BE_\nu}\frac{1}{s-M_B{}^2}$$

$$\times \{|F_1(t)|^2[(s-u)^2-(m_l{}^4-t^2)-\Delta^2\Sigma^2+(\Delta^2-\Sigma^2)(m_l{}^2-t)]$$

$$+|F_2(t)|^2[2\Delta\Sigma m_l{}^2(s-u)-t(s-u)^2-\Sigma^2(m_l{}^4-t^2)-\Delta^2\Sigma^2 t$$

$$+t(\Delta^2-t)(m_l{}^2-t)]$$

$$+|F_3(t)|^2 m_l{}^2(m_l{}^2-t)(\Sigma^2-t)$$

$$+2\,\mathrm{Re}\,F_1(t)F_2{}^*(t)[\Delta^2\Sigma(m_l{}^2-2t)+\Delta m_l{}^2(s-u)-\Sigma(m_l{}^2-t)(m_l{}^2+2t)]$$

$$+2\,\mathrm{Re}\,F_1(t)F_3{}^*(t)[\Delta m_l{}^2(\Sigma^2-t)-m_l{}^2\Sigma(s-u)+\Delta m_l{}^4]$$

$$+2\,\mathrm{Re}\,F_2(t)F_3{}^*(t)[\Delta\Sigma m_l{}^4-m_l{}^2t(s-u)]$$

$$+|G_1(t)|^2[(s-u)^2-(m_l{}^4-t^2)-\Delta^2\Sigma^2+(\Sigma^2-\Delta^2)(m_l{}^2-t)]$$

$$+|G_2(t)|^2 m_l{}^2(m_l{}^2-t)(\Delta^2-t)$$

$$+|G_3(t)|^2[2\Delta\Sigma m_l{}^2(s-u)-t(s-u)^2-\Delta^2(m_l{}^4-t^2)-\Delta^2\Sigma^2t+t(m_l{}^2-t)(\Sigma^2-t)]$$

$$+2\,\mathrm{Re}G_1(t)G_2{}^*(t)[m_l{}^2\Delta(s-u)-\Sigma m_l{}^2(\Delta^2-t)-\Sigma m_l{}^4]$$

$$+2\,\mathrm{Re}\,G_1(t)G_3{}^*(t)[\Delta(m_l{}^2-t)(m_l{}^2+2t)-\Sigma m_l{}^2(s-u)-\Sigma^2\Delta(m_l{}^2-2t)]$$

$$+2\,\mathrm{Re}\,G_2(t)G_3{}^*(t)[\Delta\Sigma m_l{}^4-m_l{}^2t(s-u)]$$

$$+2\,\mathrm{Re}\,F_1(t)G_1{}^*(t)\cdot 2[m_l{}^2\Delta\Sigma-t(s-u)]$$

$$+2\,\mathrm{Re}\,F_1(t)G_3{}^*(t)\cdot 2\Delta[t(s-u)-m_l{}^2\Delta\Sigma]$$

$$+2\,\mathrm{Re}\,F_2(t)G_1{}^*(t)\cdot 2\Sigma[m_l{}^2\Delta\Sigma-t(s-u)]$$

$$+2\,\mathrm{Re}\,F_2(t)G_3{}^*(t)\cdot 2\Sigma\Delta[t(s-u)-m_l{}^2\Sigma\Delta]\}$$

where

$$\Sigma = M_B + M_{B'}$$

$$\Delta = M_B - M_{B'}$$

$$a = \begin{cases}\cos\vartheta \\ \sin\vartheta\end{cases} \quad \text{for} \quad |\Delta Y| = \begin{cases}0\\1\end{cases}$$

XV. Deep Inelastic Neutrino Scattering

1. Kinematics

Processes under consideration are

$$\nu_l(q_1) + N(p) \to l^-(q_2) + \text{anything}$$

$$\bar{\nu}_l(q_1) + N(p) \to l^+(q_2) + \text{anything}$$

$$q_1 + p = q_2 + p_n$$

$$q = q_1 - q_2$$

In the Laboratory system:

$$q_1{}^0 = E$$

$$q_2{}^0 = E'$$

$$\measuredangle(\mathbf{q}_1, \mathbf{q}_2) = \vartheta$$

In the following, the lepton mass is neglected.

$$q^2 = -4EE' \sin^2 \frac{\vartheta}{2}$$

Define also

$$\nu = \frac{(pq)}{M} \qquad M = \text{nucleon mass}$$

$$\nu_{\text{Lab}} = E - E'$$

$$p_n{}^2 = W^2$$

$$\nu = \frac{W^2 - M^2 - q^2}{2M}$$

2. Laboratory Cross-Section and Structure Functions

$$\frac{d^2\sigma^{\nu,\bar{\nu}}}{dE'\,d\Omega} = \frac{G^2}{4\pi^2}\frac{E'}{E}\frac{1}{M} W^{\mu\nu}\{q_{1\mu}q_{2\nu} + q_{1\nu}q_{2\mu} + \tfrac{1}{2}g_{\mu\nu}q^2 \mp i\varepsilon_{\mu\alpha\nu\beta}q_1{}^\alpha q_2{}^\beta\}$$

Upper signs refer to ν and lower signs to $\bar{\nu}$. M is the nucleon mass.

$$W_{\mu\nu} = (2\pi)^6 \frac{M}{2} \sum_n \langle p|j_\mu(0)|p_n\rangle\langle p_n|j_\nu^\dagger(0)|p\rangle\delta^{(4)}(p_n - p - q)$$

$$= 2\pi^2 \int d^4x \exp(iqx) \langle p|[j_\mu(x), j_\nu^\dagger(0)]|p\rangle$$

j_μ is the weak hadronic current which may be written in the quark model

$$j_\mu^\dagger = \bar{p}\gamma_\mu(1 + \gamma_5)(n \cos \vartheta + \lambda \sin \vartheta)$$

$$W_{\mu\nu}^*(p, q) = W_{\nu\mu}(p, q)$$

The covariant decomposition is

$$W_{\mu\sigma}(p, q) = -\left(g_{\mu\sigma} - \frac{q_\mu q_\sigma}{q^2}\right) W_1(q^2, \nu) + \left(p_\mu - q_\mu \frac{M\nu}{q^2}\right)\left(p_\sigma - q_\sigma \frac{M\nu}{q^2}\right) \frac{W_2(q^2, \nu)}{M^2}$$

$$+ i\varepsilon_{\mu\sigma\alpha\beta} q^\alpha p^\beta \frac{W_3(q^2, \nu)}{2M^2} + q_\mu q_\sigma \frac{W_4(q^2, \nu)}{M^2}$$

$$+ (p_\mu q_\sigma + q_\mu p_\sigma) \frac{W_5(q^2, \nu)}{M^2}$$

The structure functions W_i are dimensionless.

$$W_i^*(q^2, \nu) = W_i(q^2, \nu) \qquad i = 1, \ldots, 5$$

The structure functions W_4 and W_5 do not contribute in the limit $m_l = 0$

$$\frac{d^2\sigma^{\nu,\bar{\nu}}}{dE'\, d\Omega} = \frac{G^2}{2\pi^2} \frac{E'^2}{M} \left\{2 \sin^2 \frac{\vartheta}{2} W_1(q^2, \nu) + \cos^2 \frac{\vartheta}{2} W_2(q^2, \nu)\right.$$

$$\left. \mp \frac{E + E'}{M} \sin^2 \frac{\vartheta}{2} W_3(q^2, \nu)\right\}$$

Define

$$Q^2 = -q^2$$

$$\frac{d^2\sigma^{\nu,\bar{\nu}}}{dQ^2\, d\nu} = \frac{G^2}{2\pi} \frac{E'}{E} \frac{1}{M} \left\{\frac{Q^2}{2EE'} W_1(Q^2, \nu) + \left(1 - \frac{Q^2}{4EE'}\right) W_2(Q^2, \nu)\right.$$

$$\left. \mp \frac{E + E'}{M} \frac{Q^2}{4EE'} W_3(Q^2, \nu)\right\}$$

The structure functions can be obtained from $W_{\mu\nu}$ by the following projections:

$$W_1 = \frac{1}{2}\left\{\frac{p^\mu p^\nu}{M^2[1 - (\nu^2/q^2)]} - g^{\mu\nu}\right\} W_{\mu\nu}$$

$$W_2 = \frac{1}{2[1 - (\nu^2/q^2)]} \left\{ \frac{3p^\mu p^\nu}{M^2[1 - (\nu^2/q^2)]} - g^{\mu\nu} \right\} W_{\mu\nu}$$

$$W_3 = \frac{i\varepsilon^{\mu\nu\sigma\lambda} q_\sigma p_\lambda}{q^2 - \nu^2} W_{\mu\nu}$$

3. Scaling Behaviour

Define

$$\omega = -\frac{2M\nu}{q^2} \qquad 1 \leqslant \omega < \infty$$

$$x = \frac{1}{\omega} \qquad 0 < x \leqslant 1$$

$$y = \frac{\nu}{E} \qquad 0 \leqslant y \leqslant 1$$

In the limit $q^2 \to -\infty$, ω fixed, the scaling behaviour is

$$W_1(q^2, \nu) \to F_1(\omega)$$

$$\frac{\nu}{M} W_2(q^2, \nu) \to F_2(\omega)$$

$$\frac{\nu}{M} W_3(q^2, \nu) \to F_3(\omega)$$

$$\frac{d^2\sigma^{\nu,\bar{\nu}}}{dx\,dy} = \frac{G^2}{\pi} EM\{xy^2F_1(x) + (1 - y)F_2(x) \mp (y - \tfrac{1}{2}y^2)xF_3(x)\}$$

The Callan-Gross relation is [10]:

$$2xF_1(x) = F_2(x)$$

4. Crossing, Isospin and Positivity Relations

$$W_{\mu\nu}{}^{(\nu)}(p, -q) = -W_{\nu\mu}{}^{(\bar{\nu})}(p, q)$$

$$W_i{}^{(\nu)}(q^2, \nu) = -W_i{}^{(\bar{\nu})}(q^2, -\nu) \qquad i = 1, 2, 3$$

$$W_i{}^{\bar{\nu}n} = W_i{}^{\nu p} \qquad i = 1, 2, 3$$

$$0 \leqslant \frac{1}{2M}\sqrt{\nu^2 - q^2}\,|W_3{}^{\nu,\bar{\nu}}| \leqslant W_1{}^{\nu,\bar{\nu}} \leqslant \left(1 - \frac{\nu^2}{q^2}\right) W_2{}^{\nu,\bar{\nu}}$$

5. Helicity Cross-Sections

Let σ_i ($\bar{\sigma}_i$) be the cross-section of a virtual positive (negative) vector boson on a proton. $i = L, R, S$ for left, right and scalar components.

$$\sigma_i = \frac{2\pi}{2M\nu + q^2} \varepsilon_i^{\mu *} W_{\mu\rho}{}^{(\nu)} \varepsilon_i^{\rho}$$

$$\bar{\sigma}_i = \frac{2\pi}{2M\nu + q^2} \varepsilon_i^{\mu} W_{\mu\rho}{}^{(\bar{\nu})} \varepsilon_i^{\rho *}$$

where

$$(\varepsilon_L) = \frac{1}{\sqrt{2}} (0, 1, -i, 0)$$

$$(\varepsilon_R) = \frac{1}{\sqrt{2}} (0, 1, +i, 0)$$

$$(\varepsilon_S) = \frac{1}{\sqrt{Q^2}} (\sqrt{\nu^2 - q^2}, 0, 0, \nu)$$

$$W_2^{\left\{\bar{\nu} \atop \nu\right\}} = \frac{M}{2\pi} \frac{|q^2|}{\nu} \frac{1}{1 - (q^2/\nu^2)} \left(1 + \frac{q^2}{2M\nu}\right) \begin{Bmatrix} 2\bar{\sigma}_S + \bar{\sigma}_R + \bar{\sigma}_L \\ 2\sigma_S + \sigma_R + \sigma_L \end{Bmatrix}$$

$$W_1^{\left\{\bar{\nu} \atop \nu\right\}} = \frac{M}{2\pi} \nu \left(1 + \frac{q^2}{2M\nu}\right) \begin{Bmatrix} \bar{\sigma}_R + \bar{\sigma}_L \\ \sigma_R + \sigma_L \end{Bmatrix}$$

$$= W_2{}^{(\nu)} \left(1 - \frac{\nu^2}{q^2}\right) \begin{Bmatrix} (\bar{R}) + (\bar{L}) \\ (R) + (L) \end{Bmatrix}$$

$$W_3^{\left\{\bar{\nu} \atop \nu\right\}} = \frac{-1}{2\pi} \nu \left(1 + \frac{q^2}{2M\nu}\right) \frac{2M^2}{\sqrt{\nu^2 - q^2}} \begin{Bmatrix} \bar{\sigma}_L - \bar{\sigma}_R \\ \sigma_R - \sigma_L \end{Bmatrix}$$

$$= - W_2{}^{(\nu)} \sqrt{1 - \frac{\nu^2}{q^2}} \frac{2M}{\sqrt{|q^2|}} \begin{Bmatrix} (\bar{L}) - (\bar{R}) \\ (R) - (L) \end{Bmatrix}$$

with

$$(R) = \frac{\sigma_R}{2\sigma_S + \sigma_R + \sigma_L} \leqslant 1$$

$$(L) = \frac{\sigma_L}{2\sigma_S + \sigma_R + \sigma_L} \leqslant 1$$

$$(\bar{R}) = \frac{\bar{\sigma}_R}{2\bar{\sigma}_S + \bar{\sigma}_R + \bar{\sigma}_L} \leqslant 1$$

$$(\bar{L}) = \frac{\bar{\sigma}_L}{2\bar{\sigma}_S + \bar{\sigma}_R + \bar{\sigma}_L} \leqslant 1$$

$$\frac{d^2\sigma^{\nu,\bar{\nu}}}{dQ^2\,d\nu} = \frac{G^2}{2M\pi}\frac{E'}{E} W_2(Q^2,\nu)\left\{1 - \frac{Q^2}{4EE'} + \frac{\nu^2+Q^2}{2EE'}\begin{bmatrix}(R)+(L)\\(\bar{R})+(\bar{L})\end{bmatrix}\right.$$

$$\left. + \sqrt{\nu^2+Q^2}\,\frac{E+E'}{2EE'}\begin{bmatrix}(L)-(R)\\(\bar{R})-(\bar{L})\end{bmatrix}\right\}$$

in the limit $E \to \infty$:

$$\lim_{E\to\infty} \frac{d^2\sigma^{\nu}}{dQ^2\,d\nu} = \frac{G^2}{2M\pi}\frac{E'}{E} W_2{}^{\nu}(Q^2,\nu)\left[1 + \frac{\nu}{E'}(L) - \frac{\nu}{E}(R)\right]$$

$$\lim_{E\to\infty} \frac{d^2\sigma^{\bar{\nu}}}{dQ^2\,d\nu} = \frac{G^2}{2M\pi}\frac{E'}{E} W_2{}^{\bar{\nu}}(Q^2,\nu)\left[1 + \frac{\nu}{E'}(\bar{R}) - \frac{\nu}{E}(\bar{L})\right]$$

6. Total Cross-Sections

$$\lim_{E\to\infty} \sigma_{\text{tot}}{}^{\nu,\bar{\nu}} = \frac{G^2M}{\pi} E \int_1^\infty d\omega \left\{\frac{F_2(\omega)}{2\omega^2} + \frac{F_1(\omega)}{3\omega^3} \mp \frac{F_3(\omega)}{3\omega^3}\right\}$$

$$= \frac{G^2M}{\pi} E \int_0^1 dx \{\tfrac{1}{3}xF_1(x) + \tfrac{1}{2}F_2(x) \mp \tfrac{1}{3}xF_3(x)\}$$

Define

$$\langle L\rangle = \frac{\int_0^1 dx\, F_2(x)(L)}{\int_0^1 dx\, F_2(x)}$$

$$\langle R\rangle = \frac{\int_0^1 dx\, F_2(x)(R)}{\int_0^1 dx\, F_2(x)}$$

$\langle\bar{L}\rangle$, $\langle\bar{R}\rangle$ in analogy. Then

$$\lim_{E\to\infty} \sigma_{\text{tot}}{}^{\nu} = \frac{G^2M}{\pi} E\{\tfrac{1}{2} + \tfrac{1}{2}\langle L\rangle - \tfrac{1}{6}\langle R\rangle\} \int_0^1 dx\, F_2{}^{\nu}(x)$$

$$\lim_{E\to\infty} \sigma_{\text{tot}}{}^{\bar{\nu}} = \frac{G^2M}{\pi} E\{\tfrac{1}{2} + \tfrac{1}{2}\langle \bar{R}\rangle - \tfrac{1}{6}\langle \bar{L}\rangle\} \int_0^1 dx\, F_2{}^{\bar{\nu}}(x)$$

For an isoscalar target

$$\langle L \rangle = \langle \bar{L} \rangle$$

$$\langle R \rangle = \langle \bar{R} \rangle$$

$$F_2^{\nu}(x) = F_2^{\bar{\nu}}(x)$$

so that

$$\frac{\sigma_{\text{tot}}^{\bar{\nu}}}{\sigma_{\text{tot}}^{\nu}} = \frac{1 + \langle R \rangle - \frac{1}{3}\langle L \rangle}{1 + \langle L \rangle - \frac{1}{3}\langle R \rangle}$$

consequently

$$\frac{1}{3} \leqslant \frac{\sigma_{\text{tot}}^{\bar{\nu}}}{\sigma_{\text{tot}}^{\nu}} \leqslant 3$$

7. Spectrum Independent Quantities

Define

$$\left\langle \frac{E'}{E} \right\rangle_{\nu,\bar{\nu}} = \frac{1}{\sigma_{\text{tot}}^{\nu,\bar{\nu}}} \int dQ^2\, d\nu\, \frac{E'}{E} \frac{d^2\sigma^{\nu,\bar{\nu}}}{dQ^2\, d\nu}$$

$$\left\langle \frac{Q^2}{2ME} \right\rangle_{\nu,\bar{\nu}} = \frac{1}{\sigma_{\text{tot}}^{\nu,\bar{\nu}}} \int dQ^2\, d\nu\, \frac{Q^2}{2ME} \frac{d^2\sigma^{\nu,\bar{\nu}}}{dQ^2\, d\nu}$$

For an isoscalar target:

$$\left\langle \frac{E'}{E} \right\rangle_{\nu} = \frac{\frac{1}{3} + \frac{1}{6}\langle L \rangle - \frac{1}{12}\langle R \rangle}{\frac{1}{2} + \frac{1}{2}\langle L \rangle - \frac{1}{6}\langle R \rangle}$$

They obey the following inequalities:

$$\frac{1}{2} \leqslant \left\langle \frac{E'}{E} \right\rangle_{\nu,\bar{\nu}} \leqslant \frac{3}{4}$$

$$0 \leqslant \left\langle \frac{Q^2}{2ME} \right\rangle_{\nu,\bar{\nu}} \leqslant \frac{1}{2}$$

8. Influence of the W-Boson

Let M_W be the mass of the (charged) intermediate boson and

$$\frac{G}{\sqrt{2}} = \frac{g^2}{M_W^2}$$

then

$$\frac{d^2\sigma^{\nu,\bar{\nu}}}{dx\,dy} = \frac{g^4}{\pi}\frac{2yME}{(2EMxy + M_W{}^2)^2}\left\{xyW_1(x, y) - \left(\tfrac{1}{2}xy - \frac{E}{M}(1-y)\right)W_2(x, y) \mp \frac{E}{2M}(2-y)xyW_3(x, y)\right\}$$

In the scaling region:

$$\frac{d^2\sigma^{\nu,\bar{\nu}}}{dx\,dy} = \frac{G^2}{2\pi}\frac{s}{(1+by)^2}\{xy^2F_1(x) + (1-y)F_2(x) \mp (y - \tfrac{1}{2}y^2)xF_3(x)\}$$

where

$$s = 2ME + 0\left(\frac{M}{E}\right)$$

$$b = \frac{sx}{M_W{}^2}$$

$$\frac{d\sigma^{\nu,\bar{\nu}}}{dx} = \frac{G^2}{2\pi}s\left\{\left[\frac{1}{b^2}\frac{2+b}{1+b} - \frac{2}{b^3}\log(1+b)\right]xF_1(x) + \left[\frac{1}{b} - \frac{1}{b^2}\log(1+b)\right]F_2(x) \pm \left[\frac{1}{2b^2}\frac{2+3b}{1+b} - \frac{1+b}{b^3}\log(1+b)\right]xF_3(x)\right\}$$

$$\sigma_{\text{tot}}{}^{\nu} = \frac{G^2}{\pi}Ms\int_0^1 dx\,F_2(x)\left\{\frac{1}{b} - \frac{1}{b^2}\log(1+b) + \left[\frac{1}{b^2}\log(1+b) - \frac{1}{b(1+b)}\right](L) + \left[\frac{2}{b^2} - \frac{2+b}{b^3}\log(1+b)\right](R)\right\}$$

$$= \frac{G^2}{\pi}Ms\int_0^1 dx\,F_2(x)\left\{\frac{1}{2} - \frac{b}{3} + (\tfrac{1}{2} - b)(L) - \left(\frac{1}{6} + \frac{b}{6}\right)(R)\right\} + 0(b^2)$$

XVI. Decay of the W-Boson

1. General Two-Body Decay

For the decay

$$W(k) \rightarrow a(q_1) + \bar{b}(q_2)$$

define the T-matrix element

$$\langle a\bar{b}|T|W\rangle = g\varepsilon_\lambda M^\lambda$$

ε_λ is the polarization vector of the W-boson.

$$M^\lambda = (2\pi)^3 \cdot 2\sqrt{E_a E_b}\, \langle a\bar{b}|j^\lambda(0)|0\rangle$$

Define the following decomposition:

$$\sum_{\text{Spins}} M_\lambda M_\nu{}^\dagger = (q_1 - q_2)_\lambda (q_1 - q_2)_\nu A + g_{\lambda\nu} B + \text{other terms}$$

then the decay rate is given by

$$\Gamma(W, a\bar{b}) = \frac{g^2}{48\pi M_W{}^3} \left\{ \frac{A}{M_W{}^2} |w(M_W{}^2, m_a{}^2, m_b{}^2)|^3 - 3Bw(M_W{}^2, m_a{}^2, m_b{}^2) \right\}$$

$$\begin{aligned} w(a, b, c) &= [a^2 + b^2 + c^2 - 2(ab + ac + bc)]^{1/2} \\ &= [(a + b - c)^2 - 4ab]^{1/2} \\ &= [a - (\sqrt{b} + \sqrt{c})^2]^{1/2}[a - (\sqrt{b} - \sqrt{c})^2]^{1/2} \\ (q_1 q_2) &= \tfrac{1}{2}(M_W{}^2 - m_a{}^2 - m_b{}^2) \end{aligned}$$

2. Decay Rates

$\boldsymbol{W \rightarrow l\bar{\nu}_l}$:

$$A = -4$$

$$B = -4(M_W{}^2 - m_l{}^2)$$

$$\Gamma(W, l\nu_l) = \frac{g^2 M_W}{12\pi} \left(2 - 3\frac{m_l{}^2}{M_W{}^2} + \frac{m_l{}^6}{M_W{}^6} \right)$$

$W \to 2\pi$:

$$A = \cos^2\vartheta |f_+^{\pi^+\pi^0}(M_W^2)|^2$$
$$B = 0$$
$$\Gamma(W, \pi\pi) = \frac{g^2 \cos^2\vartheta M_W}{48\pi} |f_+^{\pi^+\pi^0}(M_W^2)|^2 \left(1 - 4\frac{m_\pi^2}{M_W^2}\right)^{3/2}$$

$W \to K\pi$:

$$A = \sin^2\vartheta |f_+^{K\pi}(M_W^2)|^2$$
$$B = 0$$
$$\Gamma(W, K\pi) = \frac{g^2 \sin^2\vartheta M_W}{48\pi} |f_+^{K\pi}(M_W^2)|^2 \times \left[1 - \left(\frac{m_K + m_\pi}{M_W}\right)^2\right]^{3/2} \left[1 - \left(\frac{m_K - m_\pi}{M_W}\right)^2\right]^{3/2}$$

$W \to N\bar{N}$:

$$A = -\cos^2\vartheta [2F_1^2 + 2G_1^2 - 2M_W^2 F_2^2]$$
$$B = -\cos^2\vartheta [2F_1^2 M_W^2 + 2G_1^2(M_W^2 - 4m_N^2) + 8F_1F_2M_W m_N^2 + 8F_2^2 M_W^2 m_N^2]$$
$$\Gamma(W, N\bar{N}) = \frac{g^2 \cos^2\vartheta\, M_W}{24\pi} \left\{3(F_1 + 2F_2 m_N)^2 \sqrt{1 - 4\frac{m_N^2}{M_W^2}} - (F_1^2 - 2G_1^2 - M_W^2 F_2^2)\left[1 - 4\frac{m_N^2}{M_W^2}\right]^{3/2}\right\}$$

all form factors have to be taken at $q^2 = M_W^2$.

$W \to \pi\gamma$:

Because of gauge invariance, the matrix element can be written as (ε' = photon polarization vector)

$$M_\lambda = a_1[(\varepsilon' q_2)(q_1 - q_2)_\lambda - 2(q_1 q_2)\varepsilon_\lambda'] + a_2(\varepsilon' q_2)k_\lambda$$
$$A = -a_1^2 M_W^2$$
$$B = -a_1^2 (M_W^2 - m_\pi^2)^2$$
$$\Gamma(W, \pi\gamma) = \frac{g^2 a_1^2}{24\pi M_W^3}(M_W^2 - m_\pi^2)^3$$

in perturbation theory:

$$a_1 = \frac{e f_\pi \cos\vartheta}{2M_W^2}$$
$$\Gamma(W, \pi\gamma) = \frac{\alpha g^2 \cos^2\vartheta\, f_\pi^2}{24 M_W}\left(1 - \frac{m_\pi^2}{M_W^2}\right)^3$$

Appendix A

Other Forms of the Interaction

1. The Non-Local Interaction

The most general form of a non-local interaction of local V-A currents is

$$S = I - \frac{iG}{\sqrt{2}} \int d^4x\, d^4y\, J_\lambda(x) K^{\lambda\nu}(x-y) J_\nu^\dagger(y) + 0(G^2)$$

$$K_{\lambda\nu}(x) = M^4 \left\{ g_{\lambda\nu} F_1(x^2) + \frac{1}{M^2} \partial_\lambda \partial_\nu F_2(x^2) \right\}$$

$$F_i(x^2) = \frac{1}{(2\pi)^4 M^4} \int d^4k \exp(ikx)\, \tilde{F}_i(k^2) \qquad i = 1, 2$$

M is the characteristic parameter of the non-locality; it has the dimension of a mass.

The connection to IVB is

$$F_1(k^2) = F_2(k^2) = \left[\frac{k^2}{M^2} - 1 \right]^{-1}$$

$$M = M_W$$

2. The Renormalizable Scalar Theory

$$\mathscr{H}_{RST} = \sqrt{4\pi}\, f\{\bar{\psi}_e(1-\gamma_5)\psi_{E^0} B^- + \bar{\psi}_\mu(1-\gamma_5)\psi_{M^0} B^-$$
$$+ \bar{\psi}_{\nu_e}(1-\gamma_5)\psi_{E^0} B^0 + \bar{\psi}_{\nu_\mu}(1-\gamma_5)\psi_{M^0} B^0\} + \text{h.c.}$$

E^0, M^0 = heavy leptons

B^-, B^0 = intermediate scalar bosons

Reference [11]!

3. Limits on Additional *S*, *P*, *T* Interactions

The limit on additional P from P_{l2} decay is

$$\frac{C_P}{G} < 2 \cdot 10^{-3}$$

The limit on additional S and T from β-decay is [12]

$$\frac{C_S}{G} = -0.001 \pm 0.006$$

$$\frac{C_T}{G} = -0.0004 \pm 0.0003$$

Limits can be obtained from the μ-e ratio of hyperon decays:

$$\frac{\Gamma_\mu}{\Gamma_e} = r(x) + \gamma s(x)$$

$$r(x) = \tfrac{1}{2}(1 - x^2)^{3/2}(2 - 7x^2) + \frac{15}{4} x^4 \left[\log \frac{1 + \sqrt{1 - x^2}}{1 - \sqrt{1 - x^2}} - 2\sqrt{1 - x^2}\right]$$

$$s(x) = 5x \left\{\tfrac{1}{2}\sqrt{1 - x^2}\,(2 + 13x^2) - \frac{3x^2}{4}(x^2 + 4) \log \frac{1 + \sqrt{1 - x^2}}{1 - \sqrt{1 - x^2}}\right\}$$

$$\gamma = \frac{\mathrm{Re}\, C_S C_V^* - 6\,\mathrm{Re}\, C_A C_T^* + 6\delta\,\mathrm{Re}\, C_V C_T^* - \delta\,\mathrm{Re}\, C_P C_A^*}{|C_S|^2 + |C_V|^2 + 3|C_A|^2 + 12|C_T|^2}$$

$$\delta = \frac{\Delta}{\Sigma} = \frac{m_A - m_B}{m_A + m_B}$$

(Λ, p): $s(0.6) = 0.25$

(Σ^-, n): $s(0.41) = 0.6$

(Ξ^-, Λ): $s(0.51) = 0.42$

Appendix B

Comparison between Different Metrics

$x_4 = it$	$x^0 = x_0 = t$

Dirac-equation

$(\gamma_\mu \partial_\mu + m)\psi(x) = 0$	$(i\gamma_\mu \partial^\mu - m)\psi(x) = 0$
$\partial_\mu \bar{\psi}(x)\gamma_\mu - \bar{\psi}(x)m = 0$	$i\partial_\mu \bar{\psi}(x)\gamma^\mu + \bar{\psi}(x)m = 0$
$\bar{\psi} = \psi^\dagger \gamma_4$	$\bar{\psi} = \psi^\dagger \gamma_0$

Anti-commutation relations

$\gamma_\mu\gamma_\nu + \gamma_\nu\gamma_\mu = 2\delta_{\mu\nu}$	$\gamma_\mu\gamma_\nu + \gamma_\nu\gamma_\mu = 2g_{\mu\nu}$

Pseudoscalar

$\gamma_5 = \gamma_1\gamma_2\gamma_3\gamma_4$	$\gamma_5 = -i\gamma^0\gamma^1\gamma^2\gamma^3$

Plane-wave solution

$i\gamma_\mu p_\mu = \not{p}$	$\gamma_\mu p^\mu = \not{p}$
$(\not{p} + m)u = 0$	$(\not{p} - m)u = 0$
$(\not{p} - m)v = 0$	$(\not{p} + m)v = 0$

Energy projection operators

$P_\pm = \dfrac{m \mp \not{p}}{2m}$	$P_\pm = \dfrac{m \pm \not{p}}{2m}$

Spin projection operator

$S_n = \frac{1}{2}(1 + i\gamma_5\gamma_\mu w_\mu)$	$S_n = \frac{1}{2}(1 - \gamma_5\gamma_\mu w^\mu)$

References

[1] J. D. Bjorkén and S. D. Drell, Relativistic Quantum Mechanics. New York-San Francisco-Toronto-London: McGraw Hill. **1964**.

[2] M. Roos and A. Sirlin, Nucl. Phys. **B29**, 296 (1971).

[3] H. Ebenhöh, F. Eisele, H. Filthuth, W. Föhlisch, V. Hepp, E. Leitner, W. Presser, H. Schneider, T. Thouw, and G. Zech, Z. f. Physik **241**, 473 (1971).

[4] M. L. Goldberger and K. M. Watson, Collision Theory. New York: J. Wiley. **1964**.

[5] S. E. Derenzo, R. H. Hildebrand, and C. Vossler, Phys. Lett. **28B**, 401 (1968).

[6] S. Weinberg, Phys. Rev. **112**, 1375 (1958).

[7] Particle Data Group: T. A. Lasinsky, A. Barbaro-Galtieri, R. L. Kelly, A. Rittenberg, A. H. Rosenfeld, T. G. Trippe, N. Barash-Schmidt, C. Bricman, V. Chaloupka, P. Söding, and M. Roos, Rev. Mod. Phys. **45**, Nr. 2, II (1973).

[8] E. di Capua, R. Garland, L. Pondrom, and A. Strelzoff, Phys. Rev. **133**, B1333 (1964).

[9] R. Macek, A. K. Mann, W. K. McFarlane, J. B. Roberts, K. W. Rothe, C. H. West, and L. B. Auerbach, Phys. Rev. Lett. **22**, 32 (1969).

[10] C. G. Callan and D. J. Gross, Phys. Rev. Lett. **22**, 156 (1969).

[11] W. Kummer and E. Segré, Nucl. Phys. **64**, 585 (1965).

[12] W. Paul, Nucl. Phys. **A154**, 160 (1970).

[13] G. Ebel, A. Müllensiefen, H. Pilkuhn, F. Steiner, D. Wegener, M. Gourdin, C. Michael, J. L. Petersen, M. Roos, B. R. Martin, G. Oades, and J. J. de Swart, Nucl. Phys. **B33**, 317 (1971).

[14] A. Krops and H. Paul, Z. f. Phys. **267**, 129 (1974).

[15] J. Sandweiss, XVII. Int. Conf. High Energy Physics, London (1974).

Sources

1. H. PIETSCHMANN, *W*-induced anomalous magnetic moment of leptons, calculated with ξ-limiting formalism. Z. f. Physik **178**, 409 (1964).
2. H. PIETSCHMANN, Mass relations of SU(3) in a soluble ω-ϕ mixing model. Phys. Rev. **139**, B446 (1965).
3. J. S. NILSSON and H. PIETSCHMANN, Lectures on weak interactions. Univ. Virginia 1964/65.
4. H. PIETSCHMANN and J. S. NILSSON, Consistent theory of weak interactions. Phys. Rev. **142**, 1173 (1966).
5. H. PIETSCHMANN, Selected topics in current algebra, in: Lectures in theoretical high energy physics (H. H. ALY, ed.). London-New York-Sydney: J. Wiley. 1968. Attention: A factor "*i*" is wrong in the PCAC relation!
6. M. BREITENECKER, H. MORAWITZ, and H. PIETSCHMANN, Weak hadronic decays in the quark model. Z. f. Physik **204**, 345 (1967).
7. H. PIETSCHMANN, L. P. FULCHER, and J. M. EISENBERG, Radiative pion absorption and muon capture. Phys. Rev. Lett. **19**, 1259 (1967).
8. H. PIETSCHMANN, Semi-leptonic weak decay processes. Acta Phys. Austr., Suppl. **5**, p. 88. Wien-New York: Springer. 1968.
9. H. PIETSCHMANN and E. STREERUWITZ, On the μ-e ratio for semi-leptonic hyperon decays. Acta Phys. Austr. **29**, 362 (1969).
10. H. PIETSCHMANN, Introduction to the method of current algebra (Springer Tracts in Modern Physics, Vol. 50), p. 53. Berlin-Heidelberg-New York: Springer. 1969.
11. H. RUPERTSBERGER, On the decay $K_L^0 \to l^+l^-$. Z. f. Physik **228**, 332 (1969).
12. H. PIETSCHMANN, Weak interactions at small distances (Springer Tracts in Modern Physics, Vol. 52), p. 193. Berlin-Heidelberg-New York: Springer. 1970.
13. H. PIETSCHMANN, Structure effects in weak interactions, Acta Phys. Austr. Suppl. **7**, p. 588. Wien-New York: Springer. 1970.
14. H. PIETSCHMANN and H. STREMNITZER, Non-local effects in the decay of polarized muons. Nucl. Phys. **B22**, 493 (1970).
15. H. GROSSE, Deep inelastic neutrino-scattering. Acta Phys. Austr. **33**, 97 (1971).
16. W. JÄGER and H. RUPERTSBERGER, Second class currents and the μ-e ratio for semi-leptonic hyperon decays. Lett. Nuovo Cim. **2**, 421 (1971).
17. H. PIETSCHMANN, Weak interaction Hamiltonians, in: Nucleons and Weak Interactions (EMAN and TADIC, eds.). Zagreb: 1971.
18. H. GROSSE, Deep inelastic neutrino scattering. Lett. Nuovo Cim. **1**, 261 (1971).
19. H. PIETSCHMANN, Leptonic and semi-leptonic weak interactions. Acta Phys. Hungarica **31**, 153 (1972).
20. H. PIETSCHMANN and H. KÜHNELT, Non-local weak current current coupling and the decay $K_L^0 \to \mu^+\mu^-$. Nucl. Phys. **B36**, 615 (1972).
21. H. PIETSCHMANN and H. RUPERTSBERGER, Weak mesonic currents of second class. Phys. Lett. **40B**, 662 (1972).
22. H. PIETSCHMANN, H. STREMNITZER, and U. E. SCHRÖDER, Contributions of second class currents to hyperon decays. Nuovo Cim. **15A**, 21 (1973).

23. C. KAUFMAN and H. STREMNITZER, Asymmetries in semi-leptonic hyperon decay. Nucl. Phys. **B54**, 217 (1973).
24. H. PIETSCHMANN, Some topics in weak interactions. GIFT-lectures 8/72 (1972).
25. R. KÖGERLER, Introductory deep inelastics. Lectures at Chalmers Tekniska Högskola, Göteborg, 1972.
26. G. ECKER, K_{l3} form factors from current algebra. Nuovo Cim. **13A**, 291 (1973).

Druck: Dipl.-Ing. Schwarz' Erben KG, A-3910 Zwettl, NÖ.